The Regeneration Imperative
Revitalization of Built and Natural Assets

The Regeneration Imperative
Revitalization of Built and Natural Assets

Velma I. Grover
Faculty of Environmental Sciences
York University
Toronto, Ontario
Canada

William Humber
Director, Office of Eco Seneca Initiatives (OESi)
Seneca College
Toronto, ON
Canada

Gail Krantzberg
McMaster University
Hamilton, ON
Canada

CRC Press
Taylor & Francis Group
Boca Raton London New York

CRC Press is an imprint of the
Taylor & Francis Group, an **informa** business
A SCIENCE PUBLISHERS BOOK

Cover Paintings Credit: artist Brian Killin, from the upper left hand corner running clockwise. Carlton 506 streetcar, Nino's Barbershop, St. Clair winter, Dog Walkers

CRC Press
Taylor & Francis Group
6000 Broken Sound Parkway NW, Suite 300
Boca Raton, FL 33487-2742

First issued in paperback 2020

© 2016 by Taylor & Francis Group, LLC
CRC Press is an imprint of Taylor & Francis Group, an Informa business

No claim to original U.S. Government works

ISBN-13: 978-1-4822-3121-2 (hbk)
ISBN-13: 978-0-367-73749-8 (pbk)

Visit the Taylor & Francis Web site at
http://www.taylorandfrancis.com

and the CRC Press Web site at
http://www.crcpress.com

Foreword

Regenerating built and natural places is an opportunity not only for addressing our interaction with the environment (on which we depend for clean water, air and soil, amongst other aspects), but also with human created entities from infrastructure and neighborhoods to campus-like institutions and streetscapes. In my professional life I am constantly engaged with the manner in which a regeneration perspective supports a healthier environment while providing for the re-imagining and re-forming of the places in which we live.

Government policy has been written or enhanced to provide the necessary leadership for cities and towns all over North America to support the role of green technologies in regeneration. It provides many local municipalities and neighborhoods with a stimulus to venture from their aging commercial, industrial and residential land and building stock to places where people feel more welcomed, safe, renewed and rejuvenated.

These technologies include a broad range of initiatives, applications and actions that are undertaken by developers, building owners, municipalities (plus other levels of government) and industry in an attempt to improve their current state of existing buildings and infrastructure or to provide a higher level of standard for their new buildings and infrastructure. Specific examples of these technologies are alternative energies supplies such as solar, wind and biomass power, vegetated (green) roofs, reducing heat-island effect, rainwater harvesting, low-flow plumbing fixtures, highly efficient ventilation systems, certified renewable materials, green cleaning products, occupant controls and improved light for external and internal applications.

Two wonderful examples of large North American cities providing forward thinking approaches to regeneration and incorporating the use of green technology are Chicago and Toronto.

The City of Chicago and their *Sustainable Chicago 2015 Action Agenda* is a forward looking example of an innovative initiative to enhance this city's already highly-regarded image of sustainability and regeneration. This agenda has enabled that city along with their partners and stakeholders, to deliver a more livable, competitive and sustainable city by addressing innovative ideas across seven categories:

- Economic development and job creation,
- Energy efficiency and clean energy,
- Transportation,
- Water and waste water,
- Parks, open space and healthy food,
- Waste and recycling, and
- Climate change.

Several green technologies are showcased in these categories and include solar power procurement programs for Chicago residents, green infrastructure projects on public school campuses, compost wastewater treatment by products to fertilize park spaces and coal-free power procurement for all municipal facilities. The city's plan is advancing and is on track towards a successful conclusion.

Toronto initiated a 'Live Green Toronto' movement. Its aggressive downtown development included becoming the first city in North America to have a bylaw requiring and governing the construction and maintenance of vegetated (or green) roofs. Toronto's Green Bylaw is helping Toronto developers and residents to:

- Mitigate the impacts of development for storm water quality and quantity,
- Reduce the urban heat island effect,
- Reduce the city's overall temperature,
- Improve air quality, and
- Reduce energy consumption.

Both of these cities, amongst many others, are embracing green technologies to advance regeneration and attract innovative capital growth and talent for future success. Such measures also enhance the livability and quality of places, two pre-requisites for their long term sustainability. In simple terms it is the places we care about that will receive the attention, necessary for their long term viability.

In response to these initiatives new companies have emerged while existing firms have expanded operations to meet the demand for regeneration services. New employment opportunities have followed.

Many green technology applications in the built environment have been linked to the development and application of Leadership in Energy and Environmental Design (LEED®). Developed in the United States and now well established in Canada, LEED® is a nationally accepted benchmark for the design, construction and operation of high performance green buildings. LEED® was created to transform the built environment to sustainability by providing the green building industry with consistent, credible standards. Buildings that are awarded a LEED® certification incorporate leading-edge practices and foster green technologies that protect the environment and promote healthy working conditions.

Recognizing the need to embrace new neighborhood development and redevelopment of existing neighborhoods, the U.S. and Canadian Green Building Councils have established LEED® for Neighborhood Development (LEED-ND). The structure of LEED-ND is based on the themes of locations and linkages, neighborhood pattern, green infrastructure and buildings as well as innovation. Several green technologies are highlighted in this LEED® system and include several mentioned earlier as well as district heating and cooling. This tool will help new neighborhoods become great places that are more walkable, bike friendly, rely on green power and have energy efficient designs. In the process they will reduce our collective ecological impact while lowering the volume of carbon releases.

Although there have been many more regeneration and revitalization examples in the UK, Canada is catching up. However, there is a need for research on what would work in Canada and documentation of lessons learned locally. Based on their ground-breaking work in Hamilton (Ontario, Canada) through the Regeneration Institute for the Great Lakes (a partnership of McMaster University and Seneca College), Grover et al. are now sharing with a wider audience the means for making regeneration a winning strategy on many levels. I congratulate the authors in taking the initiative to prepare this publication and I am sure that it will stimulate energetic debate that leads to concrete actions to fix what is arguably one of the fields requiring innovation and attention.

Anthony (Tony) Cupido, Ph.D., P. Eng.
Chief Building and Facilities Officer
Facilities Management
Mohawk College
Hamilton, ON, Canada

Preface

In thinking about regeneration, the first thing that comes to mind is rebuilding old-infrastructure or redressing environmental degradation. However, regeneration is not just about rebuilding old infrastructure and rebuilding based on past experiences. Just as environment encompasses both natural and built environment, regeneration is also about rehabilitating, revitalizing and restoring eco-system function as well as urban excellence.

Built and natural environments vary from place to place depending on local conditions and climate, the prosperity of their human inhabitants and the political will and self-interest of public and private participants. Natural environments can range from the near-pristine (none of which can truly be said to be free of human impact) to the damaged, dying or destroyed. On the built side they range from the hallowed and respected, to those in need of repair, to those crumbling and abandoned. Regardless of their location and condition both natural and built environments demand attention either by a process of regeneration or by their consignment to the landfills of history.

We are realizing however that the latter choice, so common in the developed world, and increasingly at least as an option in the developing world, is fraught with problematic characteristics, not least being the wastefulness of such an approach, but also the inability of finding places for disposal or finding disposal methods with minimal impact on the surrounding environment.

Regeneration, a better alternative than decay and evolution into landfills of history, is concerned with an alternative vision of development in which we accept the responsibility for not simply fixing up and refreshing our natural and built places but reclaiming, restoring, and revitalizing them for long term prosperity, health, and ultimately inspiration.

We live in a world likely to grow in human population from seven to 10 billion in the second half of this century; in a world increasingly dominated by people living in cities or at the very least with urban expectations and lifestyles regardless of location. The world we live in (especially in North America) is also experiencing radical changes in terms of basic characteristics of urban neighborhoods and downtowns with downtown declines in older cities. However, on the positive side the world is informed by the necessity of enhancing the bio-capacity of everything from fresh air and water to pollination and robust soils.

Regeneration, i.e., reinforcing or bringing back to full life the built and natural features of both rural and urban worlds, is an explicit obligation, but one with an economic return from the financial and human investment perspective exceeding that from the 'disposal' economy.

As defined by the British Urban Regeneration Association:

> *"Urban regeneration is a comprehensive and integrated vision and action which leads to the resolution of urban problems and which seeks to bring about a lasting improvement in the economic, physical, social and environmental condition of an area."*

However Pricewaterhouse Cooper's perspective on regeneration is a broad one:

> *"Regeneration is about making tomorrow's world better. Delivering sustainable communities in a fast-moving commercial environment demands partnerships of commitment, trust and transparency. Partners must have common purpose, a shared understanding of the pitfalls and opportunities, appropriate governance and a lot of patience. There is no one-size fits all approach. To deliver inspirational regeneration, on time and within budget, you need to be inclusive, responsive to local and market needs, adopt a flexible approach and put in a lot of hard work up-front, particularly in ensuring commercial and financial viability."*

The Regeneration Institute for the Great Lakes (ReIGL), a partnership of McMaster University's Engineering and Public Policy Program, and Seneca College's Office of Eco-Seneca initiatives, is a laboratory of practice founded on the authority of research embedded in the modern university, and the practical acumen reflected in the contemporary college system. We are pleased to share what we have learned to date on regeneration practices and provide a foundation for further examination and action.

Velma I. Grover
William Humber
Gail Krantzberg

Contents

About the Authors

Dr. Velma I. Grover has vast experience in international development with international policy think-tank (United Nations University), consulting (UNESCO, Economic Research Forum, Egypt) and teaching at Universities (McMaster University and York University), thus developing a deep understanding of stakeholder issues and policy research (mainly in the area of water and waste). She has worked internationally on different water issues, ranging from capacity building in integrated water resources management, impact of climate change on the water cycle, trans-boundary lake basin management, integrating the supply of safe drinking water and sanitation to health and the reduction of poverty, and introducing good governance in Asia, Africa and North America. She was also invited by Oxford University to organize a half day workshop called 'The Looming Water Crisis as the Himalaya Glaciers Melt' at the Smith School World Forum on Enterprise and Environment at Oxford in July 2009. Velma has been a visiting scholar at: Kalmar University (1999), the Smith School of Enterprise and Environment at Oxford (2008), and the School of International Relations and Public Affairs, Fudan University, Shanghai China (2009). She was also a Bryant Drake Visiting Professor at Kobe College, Japan (2013–14). She was selected by an international selection committee, chaired by Queen Noor of Jordan, to attend the first leadership workshop by the United Nations University in Jordan in 1997. Velma has authored/edited 12 books and has published numerous articles in journals. She is also an Adjunct at McMaster University and York University.

Professor William Humber is Director of the 'Office of Eco Seneca initiatives' (OESi) at Seneca College in Toronto. He was formerly a Chair in the Faculty of Applied Science and Engineering Technology at Seneca College where he was responsible for the Centre for the Built Environment, one of two designated centres of excellence at the College. His lengthy 38-year career in the college system also includes work in Continuing Education and community outreach. Humber has a Masters in Environmental Studies from York University (1975). He is author of 11 books. His most recent (2009) from Dundurn Press was written in collaboration with his son Darryl. It is entitled, 'Let It Snow: Keeping Canada's Winter Sports Alive' and covers Canada's winter sports heritage and the ways climate change threatens their survival.

He has lectured on environmental and energy topics as they relate to education at conferences in the United Kingdom (The Energy in the City Conference at London South Bank University, June 2010, on the topic of Lifecycle Guarantors of Sustainability), in Europe (The Sustainable City Conference in Estonia in 2006, on the topic of The Neighbourhood Imperative), in the United States (Chairs Academy in Los Angeles in 2003, on the topic of Urban Sustainability) and throughout Canada (Association of Canadian Community Colleges on energy management training). In 2013 he was recognized with Canada Mortgage and Housing Corporation's annual award for educational leadership in sustainability for his work in creating Seneca College's Green Citizen Campaign which can be viewed at www.thegreencitizen.ca.

Dr. Gail Krantzberg is Professor and Director of the Engineering and Public Policy Program in the School of Engineering Practise at McMaster University offering Canada's first Master's Degree in Engineering and Public Policy. Gail completed her M.Sc. and Ph.D. at the University of Toronto in environmental science and freshwaters. She worked for the Ontario Ministry of Environment from 1988 to 2001, as Coordinator of Great Lakes Programs, and Senior Policy Advisor on Great Lakes. In her tenure there she was intensely engaged in binational Great Lakes science and policy venues, including direct interactions with the Great Lakes Commission, Board membership on the Georgian Bay Forever, the Great Lakes Observing system, President of the International Association of Great Lakes Research, The Great Lakes St. Lawrence Cities Initiative, Board Member of the Canadian Water Foundation, member of the International Joint Commission's Water Quality Board, Sediment Priority Action Committee, Indicators Implementation Task Force, and Council of Great Lakes Research Managers. Dr. Krantzberg was the Director of the Great Lakes Regional Office of the International Joint Commission from 2001 to 2005. In 2007 she was appointed as an adjunct faculty member of the United Nations University Institute for Water and Environmental Health and participated in the twinning of the Laurentian and African Great Lakes (principally Lake Victoria). She has authored six books and more than 140 scientific and policy articles on issues pertaining to ecosystem quality and sustainability and is a frequent speaker to media and the public.

Chapter 1

The Regeneration Imperative in a Changing World

Prelude

Prosperity!
What is it good for?

Ask some environmental advocates and they might say "absolutely nothing",[1] given its ruinous consequences for the planet, but it's certain that a majority of people treasure its attainment. Even those who intentionally, rather than by circumstance, chose a minimalist existence know they can at any time return to a life of plenty. The conundrum of our time is the way improvement in some aspects of human living contributes to decline in others. It recognizes that this decline is real, takes seriously the consequences of such decline, and engages in a mature conversation about its resolution.

Prosperity describes human success in improving quality of life from the level of education available to all and the longevity of each human life, to our having a degree of comfort and personal satisfaction unimaginable in previous times. Its achievement however rests on our continued use and disposal of resources from today's living world and also those created and stored over millions of years.

A measure of this prosperity is the United Nations' Human Development Index (HDI). A complementary indicator is the ecological and carbon (eco/carbon) footprint describing our use and disposal of resources. Ecological footprints often embed the carbon within their total calculation of all uses and impacts on nature. On the other hand the extraction, decline and disposed

---

[1] An homage to the Motown label song 'War' popularized by Edwin Starr in 1970.

impact of fossil fuels, created from transformed dead organisms, are often reported as a standalone carbon footprint. For our purposes the two are recognized as elements of a combined eco/carbon footprint.

The commonly shared interface between these two is our measure of bio-capacity. In broad terms it describes those ecological services or natural capital on which our prosperity is based, from fertile soils and insects pollinating crops to clean potable water, from wildlife balancing the ecology of places to trees sequestering carbon, cleaning the air, and storing groundwater.

Exceptions exist, but as a general rule as the combined ecological and carbon footprint of a country rises so does its improved Human Development Index, or put more simply as we use resources and turn them into market-based opportunities, the lives and comfort of citizens generally gets better (Moran et al. 2007).[2] Two items however don't improve and either go in the opposite direction or show no measureable benefit. The latter refers to civic life from judicial transparency and availability, to the democratic opportunity for all to participate in public life on a relatively equal and safe footing. Sometimes it improves; often it does not, and in many cases gets worse.

Likewise bio-capacity almost always declines as the eco/carbon footprint increases. This is particularly worrying for future generations. Nor is it simply a matter of setting aside pockets of isolated nature as distinct from human settlement, as if this were an either/or proposition. In fact the ways we design our built environment, the resources required for such and the places in which we live are often the crucial aspects of determining our draw on natural capital and our ability to regenerate its bounty.

Moving forward then, what might be a provisional but manageable heuristic with which to contemplate and undertake a re-ordering of this challenge? At the risk of being too simplistic and noting that the starting point for most countries[3] is a negative ratio, i.e., the eco/carbon footprint exceeds their bio-capacity, let's consider those countries with a positive ratio (World Wildlife Fund 2006). Three big places fitting these criteria are Brazil, Russia and Canada. The first two however have varying degrees of civic inequality (and in Russia a lack of democratic transparency), along with a range of extreme

---

[2] Reporting 2003 figures (but trends continuing to this day) show the connection between HDI, ecological footprint, and bio-capacity. Wealthy Norway and United Arab Emirates (UAE) had respective HDIs (scored out of a maximum of 1) of 0.96 and 0.85. Norway had an ecological footprint per capita of 5.9 and a footprint to global bio-capacity ratio of 3.2, while the UAE's numbers for these categories were 11.9 and 6.5. On the other hand the diminished wealth of Bangladesh and Niger was reflected in respective HDIs of 0.52 and 0.28. Their ecological footprint per capita was respectively 0.5 and 1.1 and their footprint to global bio-capacity ratio was respectively 0.3 and 0.6 (471).

[3] Indicating that economically and civically advanced countries including Germany, the United States, and Japan had ecological deficits per person (measuring ecological footprint to bio-capacity) of respectively −2.8, −4.8, −3.6., while Russia, Brazil and Canada had reserves per person of 2.5, 7.8 and 6.9. Canada's bio-capacity per person of 14.5 as opposed to its per capita ecological footprint of 7.6 is just under 2:1 (3).

poverty (and in the Brazilian interior, reported cases of actual slavery). Only Canada has an advanced quality of civic life. Its regeneration ratio of bio-capacity versus carbon/eco footprint impact is 2:1, a net positive good news story only because of the country's large size in relation to its population.

Recognizing the tentativeness in this observation, but also the need for some starting point, a future 2:1 bio-capacity to eco/carbon footprint is as good a place as any to mount a foundation for robust policies and programs. It would apply not only to countries with a negative ratio but also Canada itself where the necessity may actually be more pronounced given its resource-based economic development model.

This requires a different view of growth in which bio-capacity enhancement and its associated economic opportunities would align with a rise in HDI, doing so alongside a diminishing but recognized increase in our collective eco and carbon footprint (we still need to produce the goods and services from which bio-capacity growth could occur, and with which, perhaps ironically, our eco/carbon footprint is reduced!).

While this might work for the overall environmental health of the planet on which we live, is a deliberate or intentional regeneration strategy also a way of improving civic life? One thing is certain today's conventional winner-takes-most economy, with residual crumbs 'trickling down' to everyone else, and its associated and increasing eco and carbon footprint at the expense of a declining bio-capacity, has only moderate success and many failures in improving civic life.

Regeneration as an aspect of a more rounded economic success story offers real promise. After all it is about the process of bringing life back; of lowering carbon emissions while adding to the stock of eco system services and resilient built places; of incorporating in all initiatives the capacity for their eventual re-purposing, re-use and natural evolution. It depends on a collaborative and engaging process which in its realization creates the conditions for a lasting and judicially transparent and democratic civic life.

This is not only about the natural world, as we have come to understand it, but also the human created built one. They are increasingly integrated in their impact on our associated lifestyles, comfort, and common fate. In adopting the regeneration imperative there is no better place to begin than with a growth founded on an always adjustable bio-capacity to eco/carbon footprint ratio of 2:1.

Examining the Emerging World

Soothsayers, investors, and sports pundits all claim to have special knowledge of the future. Such promises are generally harmless amusement with variable amounts of odds-determined likelihood. Policy makers understand that demographics are a safer, though by no means infallible, territory from which to plot likely conditions. Our challenge is adapting known and likely features to a continuing overlay of new technologies, political realities, and unanticipated surprises, while the ground is constantly shifting below us in unexpected ways.

The report, 'Global Trends 2030: Alternative Worlds' (National Intelligence Services 2012), says that by 2030 the world population will likely be 8.3 billion. As we go farther out the numbers are less certain with suggestions that the world population of just over seven billion in 2014 may reach 10 billion at some point in this century. Even assuming no health-related or geo-political catastrophes however that number could be as low as nine billion.

Such assurances (and given a range of a billion people the imprecision is notable), are based on the increasing urbanization of people's living places which historically has accounted for both increased affluence and declining birth rates (Jerram 2011).[4] Regarding urbanization throughout the world, "Today's roughly 50 percent urban population will climb to nearly 60 percent, or 4.9 billion in 2030" (National Intelligence Services 2012, p. v). It could be said parenthetically that, with the exception of isolated tribes, urban lifestyles and expectations are now or will be part of everyone's reality regardless of living place.

Alongside this will be a "growing global middle class, which constitutes a tectonic shift: for the first time, a majority of the world's population will not be impoverished, and the middle classes will be the most important social and economic sector in the vast majority of countries around the world" (National Intelligence Services 2012, p. iii). This is a particularly ironic expectation given current thinking in North America that the middle class is a threatened species, which may reflect the death knell of a sense of entitlement of continuing economic prosperity for those living in North America.

(a) The Nature of Urban Places

What then is the nature of these 'urban' places in which we will live? Will desirable levels of comfort, safety, security, and happiness be realized or frustrated? Nothing is certain and presently there is a raging debate between proponents of three apparently conflicting visions—one sees a rebirth of traditional cities, another says our future is largely suburban, and a third envisions a more prosaic return to the country and something approaching smaller town living.

(i) Return to the Traditional City

A rebirth of the traditional city seems to confirm the observation of American writer and Nobel Prize laureate William Faulkner that, "The *past* is never dead. It's not even past."

---

[4] Jerram examines the role of urbanization, the connected decline in birthrates, and the probable causes in the latter part of 19th century Europe long before the appearance of birth control pharmaceuticals.

Advocates of this perspective include Edward Glaeser, a Professor of Economics at Harvard and author of 'Triumph of the City: How Our Greatest Invention Makes Us Richer, Smarter, Greener, Healthier, and Happier' (Glaeser 2011), and Leigh Gallagher, managing editor at Fortune magazine and author of 'The End of the Suburbs' (Gallagher 2013). They are part of an enthusiastic and growing cadre of proponents for an urban renaissance led by empty nesters and young people, who for the first time in the post-war period show no inclination to own a car.

Their revitalized city vision, is a kind of belated acknowledgement that Jane Jacobs got it right so many years ago in her early 1960s seminal work, 'The Death and Life of Great American Cities', in which she critiqued an emerging urban world of freeways destroying inner city neighborhoods while the poor were warehoused in isolated high rise buildings away from the vibrant if somewhat derelict places in which they had once lived (Jacobs 1961). She wrote about the timeless value of street level retail, urban diversity of land use, and the safety implicit in streets full of people going about their daily lives.

Her ideas have been dusted off and updated by proponents of the new urbanism. Their leading spokesperson Andres Duany has argued that a return to a more traditional urban form with gradations of development along a line he describes as an urban transect, will do more to enhance environmental prospects than any number of one-off 'sustainability' fixes on individual homes in single-use, low density and car-dependent 'suburban' neighborhoods.

Though there are significant differences of opinion as to form and priority of traditional urban development, attributes of this approach include higher density infill within a mixed use setting, walkable and complete street options, and a retail flavor looking more like a traditional downtown than a shopping mall. These preferences go beyond the obvious locations one would expect to find them in and are increasingly part of designs for projects on greenfield lands—places once certain to become fairly traditional single-use, low density suburbs.

Some would say this is largely an upper middle class enthusiasm with few benefits for a majority of the middle class who can't afford the cost of such places. Nor does it have much appeal for a striving lower or underclass in the developed world for whom a better means of reinvigorating communities and providing a possible avenue out of poverty might be closer to what Mike Davis calls Magical Urbanism, with its equal measure of small enterprises at odds with rigorous zoning by-laws and problematic elements of 'the hustle' (Davis 2001). In the developing world the latter characterizes favelas and squatter settlements built in what are often toxic or unsavory parts of big cities. Regardless of place, all of the above might be thought of as the way traditional cities have naturally evolved over the years.

(ii) A Suburban Future

Caution is advised by Joel Kotkin in accepting such a city-centered perspective. A leading observer of the dominant suburban form in place since the end of the Second World War[5] he confidently declares, "…the suburbanization of America is likely to continue over the next decade. The 2010 Census—by far the most accurate recent accounting—showed that over 90 percent of all metropolitan growth over the past decade took place in the suburbs. Some central cities, notably New York, enjoyed decent population growth, but their increases were still below the national average."

Likewise a Canadian study led by David Gordon, director of the School of Urban and Regional Planning at Queen's University in Kingston, Ontario documented an even greater pace of suburbanization north of the American border. "Close to 95 percent of recent growth in Canada's 33 metropolitan areas took place in the suburbs, as 1.5 million Canadians moved into these areas between 2006 and 2011. In contrast, the cores of cities grew by only 90,000 people, despite the condo booms in Toronto and Vancouver" (Cook 2013).

Canada which only mildly suffered the consequences of the 2008 Recession may thus be a proxy for what would have been an even higher suburbanization rate in the United States had its markets, particularly those in real property, not taken such a direct hit.

While suburbs remain the content of both cultural and contemporary urbanist critique, an affectionate portrait of such places is found in D.J. Waldie's appropriately named 'Holy Land', an account of a prototypical post-Second World War suburb of tract houses in Lakewood, California (Waldie 1995). It's a reminder of the power of such places to meet homeowner dreams while creating special moments of personal and family identity. More recently Benjamin Ross in 'Dead End' recognizes the way suburbs, supported by the legal system, fulfill the social striving of many of its residents (Ross 2014).

(iii) Disbursed Places

Rebelling against either of the above two possibilities is a third way summarized in James Howard Kunstler's essay, 'Back to the Future: A road map for tomorrow's cities'. He writes, "I see our cities getting smaller and denser, with fewer people" (Kunstler 2011). He calls suburbia a fiasco given its almost total dependency on great quantities of cheap oil. Nor is he a fan of downtown skyscrapers dependent on air conditioning and cheap electricity. Residential living in the inner city has taken the place of productive enterprises such as manufacturing while the water-borne trade of nearby lakes, rivers and oceans has been replaced near the shoreline by jazz festivals and posh restaurants.

---

[5] http://www.forbes.com/sites/joelkotkin/2012/07/31/americas-future-is-taking-shape-in-the-suburbs/.

His critique is not without its merits. Big downtown law firms in cities like Vancouver struggle to justify their location when the buildings in which they're located have higher revenue potential as either apartments or condos. As well the individual trying to purchase a home in a successful city must often overcome a bidding war in which the original asking price is pegged at a rate guaranteed to drive the final cost beyond the means of most two income families. Given such market conditions it would hardly be surprising that businesses and individuals would find smaller, less populated places more attractive.

Small town living after all not only has a remaining, though threatened, charm, but its homes can be purchased, or its offices rented, the old fashioned way. One submits a bid below the asking price and haggles over an agreed meeting point somewhere in the middle. Those who imagine the tranquil pleasures of such a lifestyle talk about a world in which one knows one's neighbors, and where everything is environmentally pristine. Cynics however might note that knowing one's neighbors, or them knowing you too well, can be oppressive, while the green image of such places has been soundly refuted by David Owen in his analysis of the environmental benefits associated with a Manhattan residence versus that of the commuter from Connecticut (Owen 2009).

(b) Planetary Cities and Lifestyles in a World of Megapolitan Regions

These conversations about urban form continue a debate with roots in the rival points of view once espoused by two respected mid-20th century urban thinkers. Le Corbusier's prescription for high-rises plunked down amidst a sprawling network of freeways was challenged by Frank Lloyd Wright's urban paradise of a grand home on an acre of land outside the city. At least for Wright it all seemed right until the city he critiqued sprawled closer to his compound. Le Corbusier's vision on the other hand became those high-rise islands of despair criticized by Jane Jacobs. Many are being torn down, or significantly retrofitted, and replaced with more vibrant, mixed-use neighborhoods connected to the surrounding urban core such as Regent Park in Toronto.

Kunstler's critique is the most radical of the three for it envisions a world of decline brought on by the high cost of increasingly scarce energy resources in which neither the big city downtown of high-rises and condominiums, nor the sprawling car-dependent suburbs are viable. His re-envisioned urban places, though smaller, would be "compact, dense, mixed-use, and composed of neighborhoods based on the quarter-mile walk from center to edge—the so-called five-minute walk, which is a transcultural norm found everywhere in pre-automobile urban communities" (Kunstler 2011).

How such places could either co-exist with, or replace, the significant built form of cities and suburbs is unclear. Individuals have invested significant financial resources, along with personal and family identities, in existing places, as have public and private agencies dependent on the connections between people and businesses within a knowledge and service economy (Binelli 2013).

It may be however that all three of the above perspectives will be realized. Increasing urbanization, with disagreements continuing over favored built form, is a given barring unanticipated catastrophe in which case all bets are off and Kunstler's vision has sudden validity. Assuming no such 'long emergency', the more successful urban territories will continue to grow as 'planets in their own right'[6] flourishing within an increasing regional identity and overseen by often befuddled levels of government, remnants of a once vigorous and balanced rural society alongside a modestly sized city (Kunstler 2005).

These successful 'planetary' cities function in what Nelson and Lang describe as megapolitan regions and clusters in their book 'Megapolitan America' (Nelson and Lang 2011). This more real city stretches 100 miles or more from the old traditional downtown of a historically known and well-defined setting. It contains what were once somewhat independent mid-sized cities, small towns and villages, as well as more recently developed satellite-like urbanized places, or edge cities (Garreau 1992). As oft as not the latter places, while eager to assume the identity of a city, owe their recent success as much to their location within this region as with any intrinsic economic uniqueness they might have.

It is the daily lives of residents in these megapolitan places however which give grounded reality to these statistics and projections. Local and shared urban identities are now matched by a regional, existential lifestyle as citizens attend to individual pursuits and interests regardless of geography.

What began as a commuter lifestyle of work separated from one's living place, today includes multiple attributes of daily life. Learning for the majority of children now begins with transit on yellow buses rather than walking to a neighborhood school. Post-secondary institutions that had some regard for a specific urban catchment area now compete for students within the broad commuter range of the city region. Places of worship have moved from just around the block to big-box religious settings whether these be a church, mosque or synagogue.

Recreation, whether it's one's weekend golf dates, or the kids' hockey or soccer tournaments and tryout camps, now often requires travelling hundreds of miles. Entertainment, be it live theatre or cinemas with limited movie releases, attract a diverse audience of commuters. Family obligations range from grandparents living in distinct senior's residences, to grownup children buying downtown condominiums.

Even the sports teams for which one cheers are now supported by a regional audience whose allegiance can move back and forth between nearby and even distant markets depending on the success of competing teams and

---

[6] A metaphorical observation made at the first World Architecture Day event sponsored by World Architecture News 1 October 2012 in London UK.

with little regard for hometown identity. Shopping has become a lifestyle in its own right, and specialty markets, antique fairs, and mega stores threaten the very existence of once prominent main streets. All of these once local features of life are increasingly found in multiple and widely separated municipal jurisdictions.

Within these megapolitan regions aspects of the suburban model will continue. Regardless of the price of oil there is no reluctance to budget for its use. Likewise car ownership may decline but short term rentals, long term leasing, shared car ownership, car clubs, taxis and associated 'underground' web-based options, will increase. It's no wonder satellite radio, despite the preponderance of free alternatives, has become so popular when people spend so much time in their cars. Citizens have accommodated themselves to this new world faster than their politicians whose wedge issue debates seem increasingly irrelevant to most lives. In many cases the only regional governance entities are those with limited oversight of water management or public transit.

Alongside this however a vibrant urbanity has an increasing vogue and recognized value in these successful city regions, not only for its multiple public health benefits,[7] but for its sheer joy and conviviality (Florida 2002, Illich 2000). The practical features of a diverse urban realm are at least contemplated and in many cases implemented in either new developments or the retrofitting of existing ones, regardless of their location in the megapolitan region.

There is after all only so much space and opportunity for intensive and mixed urban development in city cores and there is the occasional push back of gentrification opponents. In the case of the latter it's difficult to see the advantage of rejecting change when the option is retaining a status quo of rundown, unsafe neighborhoods whose only side benefit is cheaper accommodation, which will eventually succumb to wear, tear, and collapse. It would be far better to address the challenges of the disadvantaged through social programs not dependent on a deteriorating urban environment.[8]

As we see more ambitious concepts of urban form and identity expanding in the re-design or construction of traditional suburban locales, we might expect that both Kotkin and Gallagher can claim victory (Dunham-Jones and Williamson 2008). Novelist F. Scott Fitzgerald described the test of a first-rate intelligence as "… the ability to hold two opposing ideas in mind at the same time and still retain the ability to function." Megapolitan residents may be Fitzgerald's living examples as they function in, and move back and forth

---

[7] Walkscore Professional at http://www.walkscore.com/professional/public-health-research. php documents multiple studies showing correlations between the walkability of a place as scored by tools such as walkscore.com and levels of obesity, diabetes, and other public health maladies.

[8] In the United States this issue has become associated with white middle class newcomers, captivated by the pull of traditional urban areas, replacing long established African American households, thus further muddying the waters of social discourse on this topic.

between, both traditional but updated urban settings, and contemporary but evolving suburban realms.

Kunstler's vision as well of denser, walkable, and varied neighborhoods for smaller discreet urban places fits quite comfortably into either of the two rival urban futures, and has its greatest relevance to the challenge of retrofitting suburbia for increased urbanity and variety. Even aspects of his smaller discreet places have attained a boutique quality within some city regions.

Ontario's Places to Grow land use formulation for instance is now 10 years old and is a model for study by other city regions. The plan retains large expanses of land for a more traditional countryside of rural amenity, scenic vistas, and water source protection, within which niche places can, and do, survive and prosper. Such places however are unlikely to succeed as part of a residential prescription for renewed small town and countryside living outside a dominant city region. Instead we continue to witness an emptying out of what we have traditionally called the countryside and semi-wilderness, along with a decline in their mix of villages, small towns and in some cases even modestly sized cities. Over 10 years ago Slack et al. in their study of small, rural and remote communities in Ontario described the continuing advance of this process despite the proximity of one of the few Great Lakes city regions that to this day prospers even as manufacturing production departs.

"Population growth", they said, "in the province has become increasingly concentrated in a few metropolitan areas, and particularly in the larger metropolitan areas or city regions of southern Ontario…The rest of the province [citing the Ontario Smart Growth Secretariat 2003], in contrast, exhibits overall population decline, and relatively little job growth, despite the economic boom of the late 1990s" (Slack et al. 2013). Studies such as this influenced Ontario's 'Places to Grow' approach.

More recently Nelson and Lang report that economically diverse major city regions in the United States continue to be significantly more successful in growing their populations and economic base, and this despite the prolonged recession beginning in 2008. In the 40 year period between 1970 and 2010 the percentage of the United States population living in megapolitan areas in the 48 contiguous American states grew from 59 to 63%. Nelson and Lang project this figure to grow, essentially on the same sized landmass, to 66% by 2040 (Nelson and Lang 2011).

The only significant stressor in such areas is affordable housing. Largely rural, single industry places on the other hand continue to experience persistent poverty, with children its major victims. The writers predict, "…the nation could be moving towards two Americas rather than the many that presently exist. One, megapolitan America, would have the advantages of economic scale and highly networked spaces, and the other, non-megapolitan America, would be composed mostly of smaller, isolated places, outside the reach of economically productive networks" (Nelson and Lang 2011).

Each megapolitan region of course is distinct in its own way. Great Lakes cities particularly those on the American side have experienced out migration of population as manufacturing has declined but while some of this is by older 'cold weather averse' citizens moving to warmer climes, it also reflects the increased suburbanization and population growth outside of once robust cores such as in Detroit.

In more isolated places there may be a large net loss from migration, such as the case in northern Canadian territory of Nunavut due to many native Inuit leaving the territory for improved economic opportunities, but even here this is somewhat compensated for by growth in its settlement areas with their associated urbanized lifestyles. Iqaluit for instance now numbers (according to the 2011 census) 6,699, an increase of 8.3% from the 2006 census.

Intriguingly, but within a wildly different context, the same rural to urban shift is also occurring in the developing world. It is a journey described somewhat more positively by Douglas Saunders in 'Arrival City', where those who cobble together an existence are often able to send small amounts of money back to the impoverished villages from which they came (Saunders 2010). A person in Yemen often needs only a candy floss machine or popcorn maker to make enough income to modestly support his family. Squalor may describe the living conditions but the extreme starvation of countryside locales as popularly shown on televised charitable programs, is less pronounced.

More negatively however Mike Davis's 'Planet of Slums', describes how the extremely downtrodden are exploited in such places by criminal elements drawn from those who are only slightly less disadvantaged (Davis 2006). The protection racket nature of such places is overcome not by police, for whom these are no-go areas, but only through increasing affluence.

In commenting on a process in which makeshift encampments give way to better living conditions and a road out of poverty, the 'Alternative Worlds' report matter-of-factly says, "Owing to rapid urbanization in the developing world, the volume of urban construction for housing, office space, and transport services over the next 40 years—concentrated in Asia and Africa—could roughly equal the entire volume of such construction to date in world history, creating enormous opportunities for both skilled and unskilled workers" (National Intelligence Services 2012, p. 25).

(c) Planetary Boundaries

These developments are occurring even as the Planetary Boundaries formulation of the Stockholm Resilience Centre[9] paints a gloomy picture of the human impact on many natural systems essential for human survival and comfort, from climate predictability, to freshwater quality, to productive soil, and the bio-diversity of living things from vegetation with possible medical benefit, to the planting and pollinating contributions of bird and insect life. Knock-on negative impact for already threatened Planetary Boundaries will

---

[9] http://www.stockholmresilience.org/.

remain in the continuing extraction, use and disposal of resources regardless of where people live (Fig. 1.1).

With the very real possibility of upwards of three billion more humans on earth by sometime in the second half of the 21st century, a majority of whom will live in huge regional cities and will aspire to, or obtain, a middle class lifestyle with its consumption expectations, it's hard to be overly optimistic for planetary environmental conditions. No wonder Kunstler is so pessimistic.

To date our environmental strategies, within a business-as-usual model, appear to have had only limited ability to address this conundrum. There are multiple reasons why. One are their built-in rebound effects in which efficiency

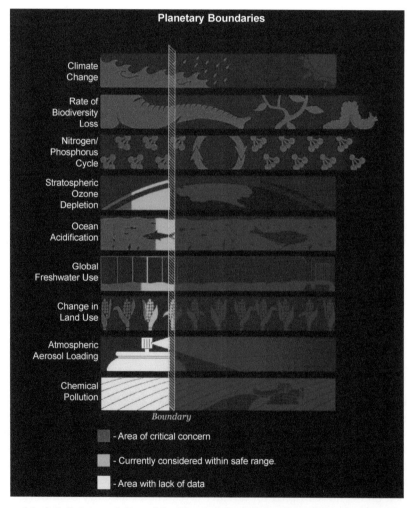

Figure 1.1. Artist's interpretation of the Planetary Boundaries formulation developed by the Stockholm Resilience Centre: provided by the Office of Eco-Seneca initiatives (OESi) at Seneca College.

measures, by reducing costs, eventually increase the absolute use of a resource and its associated waste disposal. The cumulative increase in 'per-operation reduced' CO^2 emissions has profound implications for climate instability and ocean acidification still being calculated.

Unintended consequences of even the best laid plans are also problematic. Fixing indoor temperatures at a certain level following the Fukushima nuclear plant disaster in Japan caused occupants to rely on their own inefficient heating and cooling devices, and so increase overall energy demand and emissions. Likewise measures to protect and fund increased agricultural production so as to maximize return on farming investment, often result in more pesticide-dependent planting and the tearing down of woodlands which provide necessary connections for wildlife.

Finally feel good improvement measures such as an isolated solar collector panel on a large building as a token of environmental awareness, without a concomitant energy management strategy, only marginally slows the descent into ecological trauma.

Our metrics for evaluating these measures can be problematic as well. War in the North Atlantic in the 1940s and the Deepwater Horizon oil spill in 2010 in the Gulf of Mexico, both kept fishing trawlers away from those bodies of water resulting in a significant increase in fish stocks. A prescription for war and ecological disaster is hardly a foundation for global improvement no matter how many more fish it supports. Likewise the 'world at night' view of a dark African continent while lights blaze in many parts of North America speaks not to any success in energy management in Africa but to economic distress and the failure of local infrastructure. As a speaker at the 2007 Congress for New Urbanism Conference in Philadelphia said, "The metrics of sustainability line up well with poverty."

The intent and ethics of the commonly understood notion of sustainability however are not to blame. A growth model relying on only mildly tampering with the environmental aspects of continuing resource use and its waste products and calling it sustainability, does present challenges.

Where might a new direction be found? Some have suggested a 'no growth' future or only the mild adjustment of such (Victor 2008). Others however have said we need more growth so that the resulting wealth can be used for correcting environmental maladies (Bhagwati and Panagariya 2013).

It's in the changing nature of our cities and countryside as described above however that real opportunity resides. Both allow for the direct application of measures for regenerating built and natural assets. Opportunity for instance is afforded in an increasingly abandoned countryside for the restoration and enhancement of the earth's bio-capacity features from carbon sequestering trees, to renewed soil, and waters regenerating their prior prominence and health.

As for cities, watershed damage begins to accelerate as urban impermeable surface development exceeds 10% of the land base (Beach

2002).[10] Decentralization often cited as a solution for our urban challenges would only contribute further to this calamity as greater swathes of land would pass beyond the 10% figure. It is far better to celebrate the continued concentration of urban development and populations so that always limited fiscal resources can address the resulting and inevitable degrading results of ground impermeability in these places. Bio-capacity enhancement in these growing urban centers through formulas such as net positive eco development is a longer term regeneration imperative (Birkeland 2009). At its core is moving beyond bemoaning and attempting to slow, the declining quality of our built and natural environments towards engaging with their positive re-birth and even new birth.

Another hopeful sign is the phenomenon of citizens, in what we have called the developing world, adopting a leaner consumption model. While it's one at least partly necessitated by poor infrastructure and limited resources it also encompasses the leapfrog character of new technology along with an evolving lifestyle of shared assets (Olopade 2014).[11] It contrasts with the developed world's fat consumption choices but even here 'sharing' and 'just in time' availability are appearing. Some are citing these 'lean' and 'fat' identities and trends as more realistic ways of describing our evolving world.

Accommodating these likely scenarios presents challenges across all fronts from those of housing and infrastructure, to community development and the nature of work. Lean may only go so far. Leadership awaits the 'developed' world's sense of obligation for curbing its profligacy and associated carbon and ecological impact. Building new or retrofitting the existing human-created world, within defined megapolitan regions or external territories, creates new and scaled opportunities for managing resources as part of a strategy for increasing bio-capacity, while decreasing the eco/carbon footprint.

Such a strategy recognizes that while the combined negative footprint will continue to grow due to an increasing world population and the associated cumulative increases of even 'reduced per operation impacts', measures exist to reduce it as an overall component of smart urban and economic growth.

Working in a Regeneration Economy

A regeneration strategy particularly one that hopes to achieve a 2:1 balance between bio-capacity growth versus eco/carbon footprint increase will be practiced in a world as described above which includes both increasing urbanization within successful megapolitan regions and a declining countryside in which places are downsized and single-industry economic

---

[10] P. 11 "The principle of the ten-percent impervious threshold is central to creating marine ecosystem protection programs."

[11] "Lean" societies approach consumption and production with scarcity in mind. In "fat" societies plenty is normal. In the latter gross national income is close to $50,000 per person.

prospects are increasingly marginalized. In the developing world it may include a growing middle class with associated aspirations, while in the developed world changes in different economic sectors point towards profound upheavals in work and employment.

Shifts in the latter places are an obvious concern to displaced workers as well as future generations but aren't new. Agricultural productivity for instance has expanded massively beyond its 19th century economic primacy due to new techniques, technologies, and work processes, but its labor force has shrunk in a seemingly direct proportion. Alongside this is the changing nature of agriculture itself which bears only scant resemblance to our outmoded, idealized, and pastoral vision of the 19th century family farm and its mixed rotation of crops with animals lazily dawdling in the fields.

About half the land farmed in Ontario is corn for industrial products used for sweeteners and ethanol, potatoes for chips and other snack foods, and soya for cattle who will be tomorrow's fast food hamburgers. These cash crops provide the best monetary return and in the last 20 years have led to even greater land-use intensification. Soil is depleted, pesticide use intensifies, and wooded areas are removed in order to allow for more planting.

The manufacturing sector's future is also uncertain. It might be assumed to have decamped for China and other places with low labor costs, but as Adam Davidson wrote in 'Empire of the In-Between', about the death and life of the industrial corridor linking New York and Washington "… U.S. manufacturing has never been stronger" (Davidson 2012).

He goes on to say, "While there are no universally accepted numbers, the United Nations Statistics Division calculates that the dollar value of goods made in America is at an all-time high of $1.9 trillion, just about even with China." But according to Davidson there is a significant catch and it's the same one experienced in agriculture over the last century. "The number of American workers needed to create that value has dropped steadily. There are the same number of manufacturing jobs nationwide as there were in 1941, when the country was just more than one-third its current population" (Davidson 2012, p. 32).

So while these industries may be as robust as ever, each requires far fewer workers, both skilled and unskilled. No one feels particularly secure any longer. Technology renders work less reliant on human labor, even as the digital realm makes the location of work less dependent on geography. Medical diagnoses can be performed at a distance; legal briefs can be prepared by those with access to precedent, regardless of where they work; even schools and colleges, until now one of the most distributed fields may fall victim to a standardization implicit in an on-line centralized and accrediting agency though to date no one has quite figured out how to do this. The power of once rampant labor organizations has been compromised as is their ability to control even local means of production.

Growth in the digital economy for instance has been robust but generally the number of direct employees is small relative to the size of the business. The related entrepreneurial opportunities emerging whether these are private cars for temporary hire (through a private hire-car booking app Uber), or rooms in private homes rented out to tourists (Airbnb is the world leader in travel rentals) operate in a quasi gray market outside taken for granted health and safety, insurance and taxation regimes.

This digital economy seemingly makes a centralized, one stop agency possible such as an Amazon for books and music (particularly for those who prefer tangible products) and eBay for second hand and collectible items, but it also allows for a more dispersed system of once highly centralized basic services such as energy, water and waste. In the case of physical infrastructure, design elements and communication systems for advancing both human comfort and health can be increasingly aligned with the unique physical, climatic and delivery challenges of a specific place. At the turn of this century, Sebastian Moffett predicted such a new world in his study of localized green and hybrid infrastructure (Moffett 2001). What might have been missing at the time was the proven technology to create those bespoke[12] tools necessary to meet distinct place-based peculiarities. These are now becoming possible.

Some solutions may reside in the design of practical localized manufactured products allowing for reduced environmental impact and more efficient delivery of the necessities of water and energy, as well as the management of waste. The changing nature of manufacturing is opening multiple opportunities for those with imagination and skill. At the upper end of the process is a realization that just as the off-shoring of production meant lower costs for labor it also resulted in lost knowledge. Companies sending their production overseas no longer had any idea how their products were made or of what they were constituted. So when General Electric (GE) returned some of its production to the United States, it was partially because of the uncertain quality of what they were receiving, but also because "…the product life cycle is speeding up—many goods get outflanked by 'smarter' versions every couple of years, or faster" (Fishman 2012).

GE overhauled the production of its water heater and other products by putting designers, engineers, marketers, and assembly-line workers in the same room. Realizing they no longer comprehended how their products functioned internally or of what they were made, they disassembled and then refashioned them with fewer components and more efficient operations. In the process, "The team cut the work hours necessary to assemble the water heater from 10 hours in China to two hours in Louisville" (Fishman 2012, p. 49). By so doing they were shortening and accelerating innovation cycles, reducing the required number of components in their products, and reclaiming lost knowledge.

They were also refashioning the outmoded 'miners with dirty faces era' understanding of labor-management feuds into a powerful co-operative

---

[12] "bespoke" is originally a British term for clothing made to a buyer's specification but has expanded to include individualized and personalized items from cars to computer software.

model. For labor organizations, still singing about the travails of Joe Hill, this 'us against them' mentality may threaten them with irrelevance unless they can re-imagine a new identity and opportunity.

The dependence on products and innovations emerging from a centralized mass production process may one day be surpassed by a distributed and tailored model more in tune with a disaggregated world made possible by digital connection. Technologies emerging from 3-D printing are creating vast opportunities for small companies able to respond quickly and imaginatively to market needs. Metropolis Magazine's December 2012 issue highlighted three such entrepreneurial ventures in which "…high-end fabricators use sophisticated technology to translate designers' ideas into bespoke solutions." Specificity leads to death, one entrepreneur noted, describing his job as part computer programmer, painter and carpenter (Chang 2012).

Technology and, 'infinite competition' now allow anyone to be their own manufacturer, according to Chris Anderson in 'Makers: The New Industrial Revolution' (Anderson 2012). Describing that insight, Horatia Harrod wrote, "Anyone can design almost anything—from clothes to plastic toys—using digital software. Anyone can send those designs to factories, because machines, both big and small, speak a common programming language, G-code… The factories' computer-controlled machines can be easily retooled to custom specification. And anyone can distribute the final product online to a global market" (Harrod 2012).

Is there a setting in which the characteristics of a successful city region and the changing nature of work and production can meet?

Ironically despite the potential dispersing effects inherent in new work, other factors, including opportunities for personal and family economic betterment, the possibility for increased encounters between creators, and the opportunity to experience lifestyle features at the upper end of Maslow's actualization hierarchy, are as previously noted accelerating the growth of successful 'planetary cities' or megapolitan regions and drawing people away from a declining rural hinterland. Both megapolitan and countryside places however can benefit from the opportunities associated with the regeneration of built and natural environments.

An expanding urban culture increases the opportunities for distinct and city-specific creations[13] to set those places apart in this paradoxical world of mass production and similarity, alongside distinct creations. And while many new products are art installations or architectural enhancements, smart

---

[13] The Urban Systems Collaborative (USC) occupies a unique niche in the ecosystem of urban data. Coordinated by a group from academia and industry, including IBM Smarter Cities researchers Colin Harrison and Jurij Paraszczak, USC identifies itself as being "engaged in study, evaluation and modification of real-world information to reveal emerging patterns of urban behavior that are changing the ways that people live in cities and how these changes affect the planning, design, development, governance, and operation of cities." From: http://www.planetizen.com/node/59631.

technologies are gradually being applied to hierarchical fields such as health care. Emerging is the ability to produce body replacements sensitive to the distinct characteristics of a person's individual anatomy. "American scientists have developed a hybrid printer that prints cartilage, which could one day be implanted into injured patients to help re-grow cartilage in areas such as the joints" (Anonymous 2012).

While hierarchies are an integral part of knowledge specialization and professional protection what is possible in a world of instant information access is an ability for tiers of practitioners to be in constant back and forth communication. Just as a successful medical prognosis requires the different levels of first tier doctors to be aligned with second tier nurses, emergency responders, physiotherapists, etc., so too does the design of leading edge, sustainable buildings, communities and infrastructure require first tier architects and engineers to align their ideas with the technicians, trades, and operators who manage their creations throughout a lifecycle dating forward many decades. This second tier practitioner in the built environment is for all intents, "The life cycle guarantor of sustainability" for without their commitment and knowledge whatever is built or retrofitted will soon descend (or ascend if such a relationship is attended to) in performance to the level of its operator (Moore and Paluoja 2006).

Matching unique needs with digital communication tools make unexpected solutions part of everyday possibility. "Now millions of Africans are connected to the Internet and the outside world, and they are using such technologies to mitigate deep-seated problems such as waterborne illness, which have slowed development. For example, an innovative program in western Kenya to distribute water filters and stop the use of contaminated water involved using smart phones to monitor changes in behavior." (National Intelligence Services 2012). Absent, in such places, reliable centralized systems we are likely to see more of these appropriate-level technologies combined for effectiveness with hand-held communication devices now widely available in developing countries.[14] Ironically these may drive new approaches in the developed world as the cost, localized impact of a changing climate, and a reluctance of communities to host 'big' infrastructure, increases.

The skills to operate in this new world however may reside in a quick minded ability to create common sense responses based on solid analytical and language ability in which scale can be quickly determined, self-discipline channeled for innovative thinking, and advanced design managed to produce fabricated solutions with a strong place-based emphasis. Much like the

---

[14] Open source cellular communications will most likely produce a fundamental change in the development of these nations. Access to information is propelling a DIY (do-it-yourself) culture, providing more people with opportunities to expand beyond informal regional marketplaces, and potentially develop at a larger scale. https://www.engineeringforchange. org/news/2010/06/21/open_source_cell_phone_network_could_cut_costs_to_2_per_month. html.

individually designed human cartilage, so will local areas require solutions tailored to the peculiarity of their distinction.

In a competitive world with an expanding and striving middle class, an entitlement and self-esteem based on one's communal past or geographic location will be increasingly devalued. Entrepreneurial wit will challenge big market players, while opportunities for cross-fertilization will increase. In this new world of billions of middle class aspirants, meeting their needs and interests will create new possibilities. Carbon and ecological footprint reduction could occur by replacing products with either shared items or experience-based opportunities. Likewise as essential services are driven down to the local level more place-specific entrepreneurial, workplace and cultural solutions will emerge.

Imagination will be emphasized in all features of life but most particularly in an increasingly urban world requiring new responses by humans many of whom are only recently removed from the evolved characteristics of a traditional lifestyle.[15] In the developing world it could be just-in-time social media alerts for projects such as a Mexican designer's mobile factory, in which participants are hired on a daily basis. "With technology, everything can be democratized," Amor Muñoz said (Cave, December 2012). In the developed world it could be projects as simple as that reported by Novae Res Urbis of manufacturing bicycles in several empty stalls of a Toronto condominium, though they violate zoning bylaws (Baldassi 2012).[16]

Likewise in an increasingly emptier countryside of abandoned or declining features, and beyond the emerging megapolitan 'city', the significance of not simply preserving and managing bio-capacity but developing skills to actively restore and add to its quantity and quality will make it a setting for renewed enhancement. Natural stewards for this landscape range from First Nations to former miners and mill workers who are in position to initiate a new era of rewarded and valued management as part of daily living as well as geographic and cultural preference. Likewise we can reclaim marginal farmlands for the provision of eco-system services (as has been happening with the replanting of forests either naturally or intentionally in eastern Europe), tree planting

---

[15] An interview by Robin McKie of the Observer 6 January 2013, with Jared Diamond, author of 'The World Until Yesterday: What Can We Learn from Traditional Societies?' Diamond said, "We have virtually abandoned living in traditional societies…But this was the only way of life that humans knew for their first 6 m years on the planet. In giving it up over the past few thousand years, we have lost our vulnerability to disease and cold and wild animals, but we have also lost good ways to bring up children, look after old people, stave off diabetes and heart disease and understand the real dangers of everyday life."

[16] "Although the company is small, [Councillor Adam] Vaughan said 'the imaginative forces of small enterprise [can] create a bigger enterprise. Vaughan joked that he wouldn't name the business or the location because it 'probably isn't zoned properly.' But as he pointed out, these kinds of regulations do little to foster creative solutions."

on existing roadsides and farmland as wind breaks or natural pathways for wildlife, and even with opportunities for re-wilding natural landscapes.[17]

Within cities increased urban construction and infrastructure opportunities will challenge urban theorists, designers, engineers, on-site workers and citizens to sit together, as in the case of GE's water heater model, and envision the kinds of places that will advance city living, its necessary conviviality, and its own bio-capacity growth opportunity while at the same time diminishing its extensive carbon dependence. All of these will contribute to long term urban resilience.

The leadership opportunities in the North American countryside include rebuilding bio-capacity, by providing for the education, financing, and imaginative building of appropriate communities for those living in an increasingly emptier countryside.

Ultimately we learn that despite our attempts to plot the future, as with human nature, the triumphs and tragedies of today's world survive in some form into the future, and that the old order never completely disappears. It is overlaid however by the evolving character of some things we can anticipate but also by healthy doses of serendipity, too surprising to predict. It will be a place of comfort for those with the tools of adaptability.[18]

Building an Economy Based on the Regeneration Imperative

As global urbanization is increasingly concentrated within larger city regions, and as traditional countryside identities continue their decline from both a 19th century primacy in the developed world and a more recent prominence in developing countries, the necessity and opportunity for regenerating natural and built environments has never been more timely.

Likewise as the nature of work is evolving to encompass more bespoke production tailored to local peculiarities and manufacturing needs, the

---

[17] One example, though not necessarily what we might envision for other places, is contained in an article from the New Yorker by Elizabeth Kolbert (24 December 2012) about the rewilding movement in Europe in which declining or reclaimed rural lands (such as an example in the Netherlands) are restored to a quality consistent with pre-human contact. "Oostvaardersplassen, [is] a wilderness that was also constructed, Genesis-like, from the mud. The reserve occupies fifteen thousand almost perfectly flat acres, and biologists have stocked it with the sorts of animals that would have inhabited the region in prehistoric times, had it not at that point been underwater." http://www.newyorker.com/reporting/2012/12/24/121224fa_fact_kolbert#ixzz2HaT1NF00.

[18] 'Adaptation' (Eric Klinenberg) in the New Yorker, 7 January 2013 is an urban planning description of how cities can be 'climate-proofed.' It corroborates the conclusion of this essay, describing the need for smart-grid projects, robust social infrastructure, engineered grids, restoring wetlands and planting oyster beds, intelligent design, as well as referencing the New York Resilience System, and the need to '*pro-build*' in expectation of climate change and water impacts associated with flooding and weather-related disasters. The diversity of such projects often overwhelms the linear, one-dimensional perspective of even well-meaning environmental activists, many of whom rely on a prescribed, clichéd and too often, self-interest, definition of best practices.

possibility of connecting these transformations is emerging. Digital communication and fabrication offer the potential for local infrastructure responding to the need for resilient places and locally specific solutions—an opportunity reflecting the likelihood of urban construction throughout the world on an almost unprecedented scale, while a changing countryside is ripe for bio-capacity rebuilding. Alongside these renewed places and the associated work required is an opportunity for new kinds of consumer goods, services and experiences reducing the carbon and eco footprint impact of an expanding middle class.

Regeneration already exists in many forms and places and represents an industry-wide commitment in billions of dollars. Examples include natural enhancements adding to bio-capacity, urban development correcting the built mistakes of past years while adding to the stock of revitalized city settings for both local residents and tourists, and transformative initiatives encompassing both built resiliency and natural appeal. Successful initiatives have regeneration embedded within them so that as natural elements they survive, adjust, evolve and flourish, while as built attributes they have the capacity to be re-purposed as circumstances change over time.

Regeneration's sweet spot is the way it expands economic opportunity and work opportunities by:

- Enhancing the pleasure of place,
- Increasing bio-capacity,
- Minimizing the ecological and carbon footprint of any necessary construction, and
- Expanding the resulting lifecycle uses and experiences in what's created or retrofitted.

As important as a regeneration imperative is however there is no guarantee it will become the dominant paradigm for a new global growth strategy given a political climate more often attuned to wedge issues. Add to this the self-interest in maintaining the status quo along with an out-of-date impression of what constitutes the identity of cities and countryside, and calls for action might be muted. Assuming however an improved maturity in addressing real issues and committing to the emerging world of megapolitan polities and the necessity for bio-capacity growth, project supporters and developers would still face challenges.

These include the obvious necessity of appropriate financing and development approvals, a reasonable commitment of effort with an understanding of the potential performance of the finished project, along with the availability of human talent to undertake the project together with the public's support for an initiative from the initial proposal stage to its long term functioning.

A consideration of what constitutes regeneration therefore begins with a range of interrelated and occasionally overlapping questions, all of which in the manner of a 'wicked' problem provide direction towards a best-case answer.

(a) Regeneration-specific Questions for Implementing a Regeneration Imperative

1. Does this initiative re-connect the quality of natural or built places supporting the bio-diversity and inter-relationship of living things?

   Living things require connecting and supporting natural places in order to prosper, grow and evolve. Rural landscapes that provide only pockets of disconnected woodlots in a sea of monocultural agrarianism soon lose their bio-diversity.

   Likewise suburban gardens have been found to have more songbirds than the surrounding countryside owing to each garden's (and increasingly front lawns) haphazardly varying interpretations by individual neighbors, as well their reduced or negligible use of pesticides. Successful built environments require as much connectivity as possible to improve the walking experience, public transit access, and to disperse car traffic amongst a wider range of options. Tools such as 'walkscore.com' allow one to rate the walkability of a place, and in some cases its public transit and cycling opportunities.

   Connecting these vital fabrics of life with and outside megapolitan regions through walking trails, waterfront natural connections, and measures such as daylighting urban streams (many of which, years ago, were submerged into underground channels where they are little more than storm sewers collecting detritus before depositing it further downstream), are aspects of such an approach.

2. Does this initiative add to the stock of resilient eco-system services including those supporting carbon sequestration, climate change resiliency, clean air and water, and local food production?

   We know which features of the natural environment are more successful in sequestering carbon and cleaning dirty air. We know that a porous natural landscape allows water to filter into the ground and support increased natural fecundity while reducing the impact of flash floods. We know how to design for greater resiliency in recognition of unusual weather events and climate unpredictability. We know how to grow fresh food locally and make it available in underserved neighborhoods. We know that increased bio-diversity allows for greater resiliency in the event of unanticipated natural catastrophe.

   Finally we know the value of eco-system services from honey bees pollinating plants, to the fertile soil in which these plants are grown. We know as well that many measures such as importing certain foods might make more economic sense than growing and storing them locally. Given an always limited amount of financial and human resources to be all things to all people, some choices require us to select between possible poisons. Eco-pragmatist Stewart Brand for instance contrasts the environmental damage of pesticide and energy intensive growing regimes as against a problematic role for foods modified either genetically or through invasive grafting. He asks us to make choices (Brand 2009).

3. Does this initiative create something new, or re-new and revitalize existing, human created assets from buildings to safe and vibrant communities?

Prince Charles, founder of a trust for building community (The Prince's Foundation), has observed that sustainability is only possible when people fall in love with a place. Only then will they invest their time, limited financial resources, and imagination in not just salvaging and reclaiming but enhancing what is there. As well they will guarantee that new things are built in such a way as to preserve and continue this love and have embedded within them the opportunity for future re-use for another purpose.

What might be added to this is their lifetime neighborhood quality. Is this a place that one could, if one so chose, live their entire life because it meets their needs—schooling, medical, mobility, housing, and the joy of living—at all stages of life's journey?

4. Does this initiative actively renew or return to life a decaying or lost resource, upgrade, improve or re-design an existing asset, or both of the above?

Opponents of high-rise construction, and living, often shroud their real point of view with concerns about impact on their property values, a concern almost without merit in successful megapolitan regions. While there may be some justification in objecting to the stark contrast of tall buildings next to single family houses and a desire for a more graduated-density configuration, as oft as not this is really a proxy for an antagonism towards 'those' people, though who 'those' are is never explicitly said.

'Those' people however are the focus of Toronto's Tower Renewal Project. This City of Toronto initiative aims to upgrade the energy and water efficiency of Toronto's high-rise residential stock while also providing for new construction and local employment, intensifying retail and other varied surrounding uses, while supporting public transit connections and mobility variety.

People living in these buildings are the primary focus for these opportunities. Many have affection for what has gradually evolved, even though outsiders might see these towers as somewhat run-down or architecturally dull. Important networks of human connection have been created while existing local retail often has changed to meet distinct ethnic food and material needs.

The meaningfulness of these types of places to their residents was expressed by a father following 9/11 in New York when he recalled his previous advice to his children that they could always find their way home because their neighborhood included the tallest buildings in the city. What were simply a mass of high-rises to others was a distinct local place to those living there.

5. Does this initiative anticipate and incorporate (even if a new project associated with mining, housing design, etc.), the opportunity for future re-purposing? Or stated another way, is this is a cradle to cradle initiative, as opposed to a cradle to grave initiative?

At its simplest, cradle to cradle recognizes that the items we build and use should have the property of being used forever or at least until claimed by entropy. Cradle to grave on the other hand is the symbolic touchstone of a throw-away society in which we build it, use it, and then throw it away.

Cradle to cradle has bio-mimicry features in the way it replicates the constant flow of nature's material creations for different purposes. More commercially, Cradle to Cradle is a registered trademark of McDonough Braungart Design Chemistry (MBDC) consultants and a means of accrediting companies addressing this imperative. Like the LEED (Leadership in Energy and Environmental Design) system for certifying the green credentials of buildings and neighborhoods however its proprietary aspect also has a common sense and public commons-like application.

6. Do the annual energy and environmental services requirement of this completed initiative, including its ancillary implications, require one of the following:

   a) Additional energy and environmental service requirements beyond those already used by the existing asset(s),
   b) The same energy and environmental services requirement,
   c) Approach or aim for a zero sum requirement of external energy and environmental services,
   d) Add to the stock of energy and environmental services available for distribution in the broader world beyond the boundary and demands of this initiative?

This consideration goes to the heart of whether we believe we have an unhindered right to maximize our use of resources and that such use carries no health, environmental or geo-political implications. Belief in that unhindered right allows one to rationalize their profligacy as warranted, when it is in reality a thinly disguised 'warlord-like' self interest in maintaining the status quo no matter how damaging this might be to others, including one's own children.

To the extent we lean in the other direction, the manner in which we design, or retrofit places, is central to a discussion not just on how to minimize energy and environmental impact but to actively add to the stock of such resources. Janis Birkeland's net positive development ethos for built places is a starting point for considering how to add to the stock of eco-system services in the places in which we live, work and navigate, as well as the infrastructure supporting us.

7. Does this initiative add value and character to a place, while being a replicable process for other places?

    Words like charm, conviviality, serendipity, metaphysics of place, character, and lifetime support, all play into this discussion. Charm is those memorable elements that bring one back again and again to a place. Conviviality is the robustness of human contact and interplay within which everyone either is, or feels they could be, an artist. Serendipity is the unexpected opportunities and encounters out of which something good and magnificent emerges.

    The metaphysics of place are all those undefined highlights of memory and experience from lonesome train whistles in the distance, to the wind rustling through the trees, to the music of songbirds, to the light of the late afternoon sun, to the smell of fresh cooking, to the sound of many feet on the street guaranteeing one's safety but also directing one to their destination.

    Character is what an area is known for and how it enhances those wonderful peculiarities while erasing the harmful or blasé. Lifetime support is how an area meets the needs of most people regardless of their age and mobility. Combined they are the stuff of the model place, though every place is unique. Other places may learn from them but ultimately each must respond to their own distinction.

8. By undertaking this initiative at this place does the initiative ensure that no natural or built feature is lost/sacrificed/relocated?

    While this isn't always possible it is a worthy foundation on which to continue a place's further evolution. Kevin Lynch describes the significance of well-known nodal points and places that provide ongoing identification for residents and visitors. Preservation of a once prominent but now down on its heels structure provides opportunities for envisioning renewed uses and strategic additions.

    Consider as one example New York City's High Line walkway on a once functioning overhead rail line. It was re-purposed as an in-the-sky linear greenway. Its success has seen the flourishing of both surrounding property values and retail variety, while its impact on tourism, public safety, and local pride is incalculable.

    Houston's Buffalo Bayou and Singapore's Bishan Park, wandering through their respective city centers, were revitalized and integrated with physical and recreational infrastructure. They exemplify built and natural initiatives throughout the world each distinct to their setting. Toronto's Distillery District on the site of a once prominent whisky manufacturing plant and Birmingham's Custard Factory (United Kingdom) as a creative, digital and media quarter, are other successful reimaginings of built places.

9. Does this initiative ensure that minimal embedded energy is lost on the site on which it is undertaken?

    This may seem to be of minimal concern but not only does it mean less is being transported to waste gathering places, but that the embedded energy, and therefore the work and lives of its former occupants, is honored.

    Energy resources are not only subject to varying finite or accessibility limitations, but require continual investment of always limited financial resources that could be spent on projects with increased bio-diversity benefits, and others that lower energy demand. Attending to this point not only carries into the future the past investment in securing these resources for the entire construction and maintenance process, but it limits in intensity and impact the necessity for continued extraction and use of further resources and their eventual disposal, depending on type, as carbon emissions or as nuclear waste.

10. Does this initiative have a resilient, long term authentic public engagement process which includes visioning, active participation and the assumption of obligations beyond one's self-interest?

    This is a way of tapping into local enthusiasm, project support, and the uncovering of past knowledge about a place that can transform an entire initiative. It is also about public education and the sense of obligation we all have for not only our place in the world today but that of future generations. It is the ethic of planting a tree whose full expanse one will never see or experience. It is a way of living with rather than against the world.

    It is a way of adding to rather than subtracting from the health-giving properties of such places. It is an ethic that looks beyond one's narrow self-interest and sees in a larger public realm, not only renewed places and joyous experiences, but an increased bio-diversity in which one's own individual and family life flourishes and prospers. It accelerates a process of engaged, committed and authentic civic life.

Conclusion

None of the above takes the place, or removes the obligation, of the daily requirements for attending to water and air quality monitoring and a multitude of other environmental indicators and standards. To some extent however these have passed into the quotidian and utilitarian 'taken-for-granted' aspects of today's world. They function in similar ways to how traffic signs and licensing do within policing obligations. Yet while both provide commonly understood reliability neither ultimately fends off the worst excesses of public and private behavior. The legal system of courtrooms and penalties, up to imprisonment, operates for those whose intolerable actions on the highway require public reprobation.

In the case of the environment the need is to address the big picture stuff emerging from the decisions of millions of individuals and a more discreet number of corporate entities, who in many cases are locked in an enabling partnership (we may criticize the oil sands promoters but few live without a dependence on fossil fuels).

We may not be able to answer all of the above 10 questions with certainty, recognizing that some may not be relevant while others may overlap. They are guidelines nevertheless for what really matters allowing us to determine a project's ability to meet the essential requirement of regeneration.

Over time with refinement and the development of appropriate metrics we may be able to rate their regeneration performance with greater accuracy. We could then measure the ways in which any ecological and carbon footprint increase associated with new and retrofitted activity is matched by a cumulative twofold increase in bio-capacity in our built and natural environments. Or we could determine the ways in which certain eco/carbon generating measures are ultimately approaches lowering the cumulative footprint. In this way they could meet the 2:1 ratio target for regeneration by means of fewer explicit bio-capacity measures because of the implicit bio-capacity enhancement associated with an absolute reduction in the eco/carbon footprint.

References

Anderson, C. 2012. Makers: The New Industrial Revolution. Random House, London.

Anonymous. 22 November 2012. Scientists Develop 3D Tissue Printer that Prints Cartilage. Toronto Star.

Baldassi, J. 21 December 2012. Planning refinements needed. Novae Res. Urbis.

Beach, D. 2002. Coastal Sprawl: The Effects of Urban Design on Aquatic Ecosystems in the United States, prepared for the Pew Oceans Commission.

Bhagwati, J. and A. Panagariya. 2013. Why Growth Matters: How Economic Growth in India Reduced Poverty and the Lessons for Other Developing Countries. Public Affairs, New York.

Binelli, M. 2013. Detroit City is the Place to Be. Picador, New York.

Birkeland, J. 2009. Positive Development: From Vicious Circles to Virtuous Cycles through Built Environment Design. Earthscan, London.

Brand, S. 2009. Whole Earth Discipline: An Ecopragmatist Manifesto. Viking, New York.

Cave, D. December 2012. A Factory on Bicycle Wheels, Carrying Hope of a Better Life. New York Times.

Chang, J. December 2012. Machine Histories. Metropolis Magazine, New York, NY.

Cook, M. 6 September 2013. Canada's New Identity: A Suburban Nation. Postmedia News, Toronto, Canada.

Davidson, A. 2 November 2012. Empire of the In-Between: The death and life of the industrial corridor linking New York and Washington. New York Times Magazine.

Davis, M. 2001. Magical Urbanism: Latinos Reinvent the U.S. City. Verso, London.

Davis, M. 2006. Planet of Slums. Verso, London.

Dunham-Jones, E. and J. Williamson. 2008. Retrofitting Suburbia: Urban Design Solutions for Redesigning Suburbs. Wiley, Hoboken.

Fishman, C. December 2012. The Insourcing Boom. The Atlantic Magazine, Washington D.C.

Florida, R. 2002. The Rise of the Creative Class. Basic Books, New York.

Gallagher, L. 2013. The End of the Suburbs: Where the American Dream is Moving. Portfolio/ Penguin, New York.

Garreau, J. 1992. Edge Cities: Life on the New Frontier. Anchor Books, New York.

Glaeser, E. 2011. Triumph of the City: How Our Greatest Invention Makes Us Richer, Smarter, Greener, Healthier, and Happier. Farrar, Straus & Giroux, New York.

Global Trends 2030: Alternative Worlds, a publication of the National Intelligence Council, December 2012.

Harrod, H. 7 October 2012. A Digital Guru's Plan to Save the Economy: Do It Yourself, from Seven, a publication of The Evening Telegraph.

Illich, I. 2000. Tools for Conviviality. Marion Boyars Publishers, London.

Jacobs, J. 1961. The Death and Life of Great American Cities. Random House, New York.

Jerram, L. 2011. Streetlife: The Untold History of Europe's Twentieth Century. Oxford University Press.

Kunstler, J.H. 2005. The Long Emergency: Surviving the End of Oil, Climate Change, and Other Converging Catastrophes of the Twenty-First Century. Grove Press, New York.

Kunstler, J.H. July/August 2011. Back to the Future: A Roadmap for Tomorrow's Cities. Orion Magazine, Great Barrington.

Moffett, S. May 2001. A Guide to Green Infrastructure for Canadian Municipalities. FCM Centre for Sustainable Community Development and the Sheltair Group.

Moore, S. and R. Paluoja. 2006. Life Cycle Guarantors of Sustainability from The Sustainable City IV: Urban Regeneration and Sustainability. WIT Press, Southampton.

Moran, D.D., M. Wackernagel, J.A. Kitzes, S.H. Goldfinger and A. Boutaud. 2007. Measuring sustainable development—nation by nation. Ecological Economics 64(3): 470–474.

Nelson, A. and R. Lang. 2011. Megapolitan America: A New Vision for Understanding America's Metropolitan Geography. APA Planners Press, Chicago.

Olopade, D. 28 February 2014. The End of the 'Developing World. Op-Ed contributor. New York Times.

Owen, D. 2009. Green Metropolis. Riverhead, New York.

Ross, B. 2014. Dead End: Suburban Sprawl and the Rebirth of American Urbanism. Oxford University Press, New York.

Saunders, D. 2010. Arrival City: The Final Migration and Our Next World. Knopf Canada, Toronto.

Tachieva, G. 2010. Sprawl Repair Manual. Island Press, Washington.

Victor, P. 2008. Managing Without Growth: Slower by Design, Not Disaster. Edward Elgar, Cheltenham.

Waldie, D.J. 1995. Holy Land: A Suburban Memoir. W.W. Norton, New York.

World Wildlife Fund et al. 2006. Living Planet Report. Retrieved Aug 5, 2015 from http://wwf.panda.org/about_our_earth/all_publications/living_planet_report/living_planet_report_timeline/lp_2006/.

Example of a Green Building

The Engineering Technology Building (ETB) – McMaster University

The vision of the ETB was to embrace three key objectives; to create a new entrance to the Main Street face of the campus; to provide an exemplary learning tool for the engineering students; and to demonstrate leadership with the best in sustainable design and construction practice. The ETB presents an exceptional image that is open, innovative and engaging to the community of Hamilton and beyond.

Location: Hamilton, Ontario
Size: 11,671 m² or 125,630 ft²
Certification: LEED Canada Gold

Green Technologies and Features

Construction Impact

- Over 97% of the construction waste was diverted to landfill to recycling centres.
- Erosion and sedimentation control measures were successfully implemented.

Materials

- The building has triple glazing with glass that is low E, Argon gas filled with reflective coatings between panes.
- The roof system is an R-30, non-vented substrate (NVS) light weight concrete. The roof membrane material is high albedo that reflects sunlight and heat to reduce the urban island effect.
- All the concrete in the structure incorporates residual slag from Hamilton Steel mills. This accomplishes two things; a waste product is kept out of the landfill and secondly, the slag takes the place of Portland cement. Over 300 tonnes of CO^2 was saved using slag I the concrete at ETB.

Ventilation

- The ETB features a "hybrid" heating, cooling and ventilation system, using 100% "once-through" fresh air for ventilation and heat recovery at point of exhaust, combining energy efficiency and superior indoor air quality. An enthalpy wheel reclaims the heat/cool energy in the discharged air and reconditions the new fresh air being brought in. Carbon dioxide monitors ensure air quality is maintained even with full room occupancy.
- ETB features two indoor garden spaces that function as social and academic interaction spaces while further enhancing indoor air quality through plant photosynthesis.
- Low levels of volatile organic compounds were specified in all adhesive, sealants, paints and coatings and composite wood products. All natural wood products were Forest Stewardship Council (FSC) certified.

Water Conservation

- The ETB has a unique rainwater harvesting system that collects rainfall from the roof of the building and reuses it for both non-potable and potable uses in the majority of the building. The building is licensed by the Ministry of the Environment to provide potable water through a water treatment system supplying 150 lpm to the building. The treatment system consists of multimedia filters, activated carbon filters, Ultraviolet (UV) light disinfection and sodium hypochlorite residual disinfection. Municipal water backup is provided if required. The system is designed to allow for education of students and water research. A reduction of up to 90% in municipal potable water requirements is realized.
- Low-flow plumbing fixtures and waterless urinals are in place.

Other Innovations and Technology

- ETB utilizes green power that was procured to support this industry.
- The roof system was structurally designed to host a future green, vegetated roof and solar panels.
- Efficient lighting technology utilizing T8 and T5 fixtures which are locally controlled by occupancy sensors and monitored 24/7 by the central utilities staff.
- Touch-screen kiosk in the main lobby to educate visitors and identify green features of the facility.

Picture and Description, Courtesy: Tony Cupido

Chapter 2

The Neighborhood's Role in Realizing the Regeneration Imperative

Following the events of 9/11, Mary Brendle, a former chair of Community Board 4 in Manhattan, in a letter to the New York Times (28 October 2001), provided a timely reminder about the value of the neighborhood.

> "Our collective strength as a nation and as New Yorkers," she said, "Is rooted not in power or tower, but in the love evidenced in the rescue efforts. Community breeds such devotion. Neighborhoods, not architectural statements, create bonds among people. Neighborhood residents, workers and visitors share needs, pleasures, adversities, celebrations and commemoration. Instead of replacement, we can and should expand on a trend well underway in lower Manhattan, where an influx of residential development is transforming the area from a night time desert to a vibrant neighborhood of around-the-clock activity."

The urban boundary has evolved within a megapolitan region stretching in some cases over a 100 miles from a once dominant city center. Surrounding and somewhat independent mid-size cities, small towns, rural landscapes, and more recent 'edge cities', are now places sharing all of the interests, jobs, and car dependency with a primary, though jurisdictionally defined, urban place. Governance structures at a municipal and state/provincial level are struggling to adapt to this reality, or still feigning a lack of interest (Courchene 2004).

While regeneration needs to recognize the opportunities for its fulfillment in these megapolitan regions, as well as the less populous territories beyond them, it also has a significant and possibly even more important role to play at a distinct grass roots or local level. Though increasingly urban lives are characterized by their connections to widely differing destinations for everything from work and entertainment, to religious practice and schooling, to retail variety and family obligations, at least half of one's time is spent at the local level of what we fondly call neighborhoods. They have a long and respected identification with what most of us would describe as familiar and comfortable places.

A successful regeneration process is concerned with engaging these neighborhoods, discovering their hidden assets, ideas and enthusiasms, critiquely accepting their shortcomings, addressing their public health deficiencies and correcting them, examining and realizing the opportunities for local economic initiative alongside an improved civic dimension, and designing and implementing solutions which support their contribution to a bio-capacity versus eco-/carbon footprint ratio of 2:1.

Contemporary Features of the Neighborhood

Some might consider the neighborhood idea an anachronistic and sentimental refugee from an earlier era but one now irrelevant to contemporary challenges. There are reasons for agreeing with such a conclusion. Our modern idea of the neighborhood is rooted in the rise of a property—owning middle class for whom a home has been their largest investment. Its protection and the enhancement of its value are major features of their security and wealth accumulation (Hayden 2003). Often this breeds a 'Not in My Backyard' (NIMBY) attitude among many homeowners, who often unite in neighborhood connection only when there is an external threat, such as a new development or a rise in crime. The property industry has furthered this evolution by defining a home's value as a factor of location in which neighborhood branding by use of a name and its implied identity provide its substantive meaning not only for outsiders but for those who live there.

The neighborhood idea accordingly remains confused and contested. It lacks, in most cases, any formal governance. Modern living in the developed world, and most often the desired goal of developing countries, is towards a lifestyle of consumption in which private cars go off to work, or shopping for even a jug of milk, or driving the kids to school. Public encounter is replaced by increasingly private lifestyles practiced in isolated recreation and media rooms.

Low residential densities and a lack of street connectivity have eroded the possibility of other forms of mobility including public transit, and walking to nearby places. These precious enclaves with no noxious uses, and a single use identity, bereft even of small stores and other non-harmful functions, are a product of restrictive zoning. Children and the elderly whose lives are limited by a lack of car access and the constricted radius of safe walking are marginalized residents of dead streets. The hyper connectivity of modern technology from cell phones, iPods and other electronic devices enhances the privatizing character of everyday life.

The Neighborhood Model

Despite these cautionary observations however a neighborhood as a physical entity retains three significant features. It is in varying degrees a social unit, a spatial unit, and a network of relationships and patterns of use (Chaskin 1995). Its value for policy initiatives and environmental intervention has also been recognized (Bradford 2004, Katz 2004). Diversity, tolerance, and artistic variety are key attributes for instance in the emerging idea of the creative city (Florida 2002). Cities require intriguing, differentiated mixed-use neighborhoods with retail, residential and other uses. As well the social and economic consequences of declining social capital are increasingly associated with lifestyles driven by urban sprawl, and reduced attention to neighborhood connection (Putnam 2000).

There is increasing recognition as well of the significance of place in achieving healthy lives and greater social equity (Epstein 2003). Often this is as simple as investigating the level of public services delivered in a particular neighborhood as opposed to a more affluent one. Organizations like the United Way in Toronto have defined the neighborhood as a setting for social intervention (Unity Way 2001). Regeneration strategies throughout the world focus on rebuilding local institutions and improving the quality of the urban experience through better public spaces, as well as pursuing private investment in shopping options from grocery stores to second hand shops.

(a) The Neighborhood Through Three Waves of Regeneration

The modern city is the culmination of two seemingly contradictory developments. Both of these have had, and continue to have, profound influence at a neighborhood level, hence their present discussion. The first was the ascendance of the compact urban form, often within existing and historically prescribed locations. It flourished on the back of the industrial revolution and was characterized generally by mixed uses in close proximity to each other. It was a setting in which walking was the primary means for getting about (most needs being within 400 to 800 meters of one's residence). Conditions imposed by dependence on nearby water for steam power, meant factories were often near one's backyard.

The second was the dispersing effect of later technologies that made possible the suburban character that defines so much of the contemporary city (Rae 2000). The 19th century city, notwithstanding all the idealized glory of remnant architecture and narrow streets, was for most people a miserable setting, lacking clean water and regular sanitation services. It was akin in many ways to cities in the developing world in our own time. Nor was there general agreement that something should be done about these conditions. Self-interest often argued against public expenditure.

(i) The First Wave of Urban Regeneration

Only as the level of degradation became pronounced did a class of publicly minded, privately successful entrepreneurs emerge willing to take on the challenge of civic improvement in what might be thought of as the first wave of urban regeneration. The City Beautiful movement, Garden City proponents, public health campaigners, Jane Addams' Settlement Houses and Neighborhood Centers, and Frederick Olmstead's grand park schemes, were part of this wave.

There was, however, no guarantee that such idealistic models would outlive them. One that did, ironically, was one of the last vestiges of the Progressive Movement in the United States—the escape from disease-ridden, industrially polluted inner cities to the world of the suburbs. Its associated lifestyles reflected single use zoned and low density residential areas in which walking, if not actively prevented, was discouraged by the structural layout of such places. Today these generally sedentary lifestyles are cited as a significant factor in the public health epidemics of childhood obesity, diabetes and heart disease.

The old city of the 19th century has survived, but while aspects of it remain intact, it too has been subject to policies too often favoring car dependency over walkability, downtown shopping malls turning their back on the street, or gated, similar-type, living places. Modern urban planning in post-Second World War North America was for a long time characterized by policies supporting freeways through inner city cores and the widening of roads making them more dangerous for pedestrians, particularly children and the elderly. Other approaches included the paving over of spaces, once occupied by the natural flow of streams and rivers, to rush storm water through the town and either flood those downstream or dump street refuse (from leaking car oil to dog detritus) into increasingly murky and polluted bodies of water.

(ii) The Second Wave of Urban Regeneration

The next wave of urban regeneration in the post-Second World War period saw the tearing down of once lively but 'down on their heels' mixed urban neighborhoods of homes, shops, workplaces, etc. Displaced poor people were massed in bleak high-rises on streets disconnected from the urban fabric. There were as well no nearby resources or urban variety. These neighborhoods typically devolved into dangerous no-go places for everyone except those forced by circumstance to live there.

In many cases we are still under the spell of these outdated paradigms of urban thinking that degrade built and natural environments. Why might this be?

First, it is often the general public that holds on the longest to discredited ideas of urban development, whether it be bulldozing historic inner city neighborhoods or enabling low density urban sprawl. Our human nature keeps us clinging to the familiar even if it works against our self-interest and the interests of our children, aging parents and grandchildren.

Second, we have become distrustful of new efforts promoted under the banner of urban renewal. Many people remember the damage done by large scale urban renewal efforts of past decades, and are reluctant to reopen the past. Unfortunately, our non-action only helps to maintain the status quo.

Third, many property owners oppose initiatives they sense could devalue their investments. This is easy to understand in an era when many of us rely on a bubbling real estate market to feather our nests. However, there is a downside when the public good is routinely trumped by our private interests.

Even social improvement advocates often regard the fossilized status quo of a familiar and engrained neighborhood identity as something to be preserved against any and all change. Gentrification, or the improvement of many older neighborhoods, is criticized as a cultural and economic imposition on those who per force must accept the 'tumbledown' and often disreputable character of their living place as a necessary price for less expensive housing.

With regards to the latter, James Hulme, Director of Policy and Research for the 'Prince's Foundation for Building Community', says, "the issue of gentrification is a red herring. All places are constantly changing. Without the introduction of investment, new ideas, and economic commitment, they would eventually fall into such a state of disrepair they would have to be torn done" (J. Hulme, October 2012, pers. comm.). Perhaps we should be thankful for the real estate industry's ability to upgrade areas, and leave to the realm of public policy and social service measures the fate of the dispossessed and those requiring less expensive accommodation.

Yet some things are changing in urban thinking with positive implications for neighborhoods. Real estate, for instance, is now more likely to be judged by the walkability of the surrounding neighborhood, and the choices available to meet human service and shopping needs.

(iii) The Third Wave of Urban Regeneration

The paradigms of progressive urban thinking now inform what we might call the third wave of urban regeneration. The attributes of a 'Quality Urbanism' increasingly challenge the primacy of automotive movement, big boxes on the edge of town destroying main street retail areas, or monster homes replacing streetscape integrated bungalows. Its aspirational features include the following perspectives:

Mobility equity should characterize urban places in which the car is only one of many means of getting about. Walkability should be a primary attribute, and tools such as www.walkscore.com provide a reliable score for any address in North America relative to other locations. Equity also supports the idea that 100 people on a bus should have preference in public thoroughfares to one person in a car. It means road space is allocated and designed for more than the car, most particularly public transit, bicycles, healthy pedestrians and those requiring mobility aids.

Lifetime features argue that urban places should have a diversity of ownership/tenancy options with nearby schools, medical supports, places to which a young person can walk and play safely (generally a 300 meter radius), opportunities for older people to get out and about and also rest when necessary. In other words, the kind of place where one could, if one elected to do so, live their entire life because it always meets their needs, but also inspires their imagination, at all of life's stages. This requires a range of housing, from row houses and bungalows to mid-rise apartments alongside single family dwellings. These in turn create more nearby opportunities for quality, diverse retail and improved public transit.

Safety and conviviality are somewhat paradoxically joined together in the best urban places. Safety can mean freedom from injury, assault, or risk, but also imply boredom and a failure to take chances or have new experiences. The latter are the property of at least some of the features of conviviality which describes the good feelings associated with experiencing life to its fullest, and celebrating its special moments with all due merriment with friends and neighbors. Managing that balance is the secret of special places.

Smart infrastructure describes those increasingly hybrid connections of centralized and localized water, waste and energy management and supply. It also includes nearby public institutions, as well as walkways and sidewalks seamlessly integrated alongside roadways, tending, in the best cases, to be what the Dutch call 'woonerfs' or naked streets. To the extent infrastructure delivery is localized so too are opportunities for environmental enhancements through mobilization of ecological services in design and bio-capacity improvements that advance the potential for local jobs and renewed civic spirit thorough neighborhood and regional governance.

Beauty, aesthetics and charm are value-laden aspects of quality urbanism but few disagree on their general properties. Peculiarities, we might think of as a 'metaphysics of place and spaces', connect us to a location (the sights, sounds, and smells which often have an almost indefinable, seemingly mystical quality). It is the anomalies of crooked roads, funky signs, overhanging trees, and in some cases even slightly potholed side streets which slow down or deter cars that often define the unique character and beauty of a place.

This third wave, explored in more detail later, advances a more nuanced and participative planning process, characterized by public education rather than public relations, and incorporating a wider variety of interests, voices, and points of view. Tools such as charrettes, the Prince's Foundation's Enquiry by Design (EbD), and Future search out of Australia, argue for more balance and upfront processes which prevent the costly delays and heartaches of later objection. These tools have the potential to intelligently engage the public as to the implications of their often clichéd responses to issues, as well as ways to improve outcomes.

(b) Toward a Shared Understanding

Neighborhoods however continue to be defined imprecisely despite their daily newsworthiness as part of regular commentary about a city. While everyone appreciates neighborliness, we probably disagree on what that means.

They may be the lifeblood of cities, but there are different ideas about what makes for good or inferior places. Older neighborhoods have multiple uses from residences to institutional buildings, to nearby workplaces and stores. They were fashioned on the old grid system of rectangles and smaller blocks and 90 degree angles. They can be wonderful places with plenty of history, as well as colorful citizens who may have lived there for years, and gorgeous canopies of mature trees which give many parts of each neighborhood the appearance of being in a park. They can also be places of dereliction, hopelessness, and harm to their citizens.

Regardless it is these older places that provide our best models, but also lessons, for interpreting what should characterize acts of regeneration. This is particularly relevant to the challenges of more recent, but single-use and car dependent, suburbs.

Before examining some of those features however we require a shared understanding of what a neighborhood is, a task it might be said not often spelled out in even the most self-congratulatory epistles to their value.

(i) What is a Neighborhood?

The neighborhood has been used as a jumping off point for many exercises in community planning and social policy making, but it seldom receives significant examination in its own right. Over the years however there have been some attempts to produce a standalone description.

A 1948 report, 'Planning the Neighborhood', which influenced much post-war subdivision development argued, "In a rough sense, the neighborhood as we view it corresponds rather closely to the area served by an elementary school ... . We must build not merely homes but neighborhoods if we are to build wisely for the future of America" (Solow 1948).

For the writers of this report, the neighborhood was the smallest planning unit and an area in which residents shared common services, social activities and facilities, and particularly an elementary school. Suzanne Keller's sociological treatment of neighborhoods in her 1968 study, 'The Urban Neighborhood', described a kind of 19th century inner city neighborhood slowly disappearing in the face of new suburban development. It is nevertheless a useful starting point. In defining the neighborhood she said it is,

"… a place within a larger entity, [it] has boundaries—either physical or symbolic and usually both—where streets, railway lines, or parks separate off an area and its inhabitants or where historical and social traditions make people view an area as a distinctive unit" (Keller 1968).

Her neighborhood could include geographic as well as ethnic or cultural features, psychological unity among residents who feel they belong together, and the use of the area's facilities for shopping, relaxation and schooling. Keller acknowledged that places incorporating all of these were extremely rare in modern cities. The distinguishing social characteristic of neighborhoods is the near grouping of what she calls 'proximate strangers' as opposed to friends or relatives.

More recently historian Alexander von Hoffman, referred to the term's illusiveness, calling the definition of a city neighborhood 'problematic' noting that "historians have only rarely studied the neighborhood as a distinct element in urban society." The problem he says is a tendency to accept the sociological description of the neighborhood as a homogenous unit and thus study it only as it sheds light on other matters such as ethnicity. For von Hoffman, neighborhoods have the following characteristics, "… An urban spatial unit with generally recognized geographic boundaries, a name, and some sense of psychological unity among its inhabitants" (von Hoffman 1995).

Some of the more challenging recent work on the neighborhood has come out of the architecture and planning professions as they tackle what Andres Duany of the New Urbanist movement called the issue of human habitability. A like-minded new urbanist Peter Calthorpe, describes neighborhoods as districts that are true communities of place.

"They are complex, human-scaled places that combine many of the elements of living: public, private, work, and home. They mix different kinds of people and activities in close proximity and provide places for them to interact. They provide for the everyday and sometimes random casual meetings that foster a sense of community. They create shared places that are unique to each neighborhood and shape a social geography intimately known only by those who live and work there. They are hard to design but easy to design away. And they are essential to our well-being—not just in times of crisis, but also in living our everyday lives" (Calthorpe and Fulton 2001).

For Calthorpe, neighborhoods provide society with, "Its ground-level social fabric and community identity. The neighborhood is the place where people do, or do not, gather the will, the attitudes, the resources, and the 'social capital' required to live successful lives, both locally and in the metropolitan region as a whole." As citizens, he says, we need to experience the implications of regional visions in our own neighborhoods in real life situations.

The form, density and scale of neighborhoods vary but for Calthorpe it is a walkable place with clear boundaries, and some form of identifiable center with local services and civic functions. "It includes a variety of people, offering housing opportunities for rich and poor, large family or small, young or old. Its diversity and human scale breeds a kind of intensity and sociability that creates a powerful identity and a strong sense of community" (Calthorpe and Fulton 2001).

(ii) Neighborhood Level Characteristics

Tom Scanlan author of 'Neighborhood Geography' describes a neighborhood as a comprehensible area and its internal proximity is an organizational advantage. The local neighborhood for him is a sample region providing 'small data' suitable for discussion and analysis beyond the local scale. This sense of the neighborhood as a place that provides facts whether from census information or local polling has gained in importance amongst social agencies. They are as likely to recommend local intervention strategies alongside more comprehensive regional programs. There is an argument for targeting specific actions, be these child care centers or better shopping options, at a neighborhood level rather than waiting for national programs or market place adjustments. For Scanlan, the needs of children are the primary criteria in defining a neighborhood because they are the weakest and most vulnerable.

> "Pollution and noise from traffic speeding through a residential area may be a nuisance to an adult but to a child that same traffic is more likely to be fatal. On the other hand, if a neighborhood is safe for children it is probably safe for everyone. Thus children are a prime measure of the state of our environment" (Scanlan 1984).

The neighborhood is also a very practical lifestyle and economic consideration for most people. Frederick Jarvis says that people shopping for a home first consider a neighborhood with convenience to work, shopping, and recreation; satisfactory schools and community facilities; overall quality-of-life expectations; and the individual household's position and aspirations in terms of age group, income level, social status and interests. Perhaps only after that does the question of hot tubs and recreation rooms become important, though this is debatable.

There remains a degree of intimacy either with those who live nearby or with the generalized idea of the neighborhood. Mobility may have become paramount in most people's lives, but some residential preference still follows ethnic background.

A neighborhood should not be defined by one type of physical form. It can be found in mobile trailer parks, apartment buildings, and a suburban subdivision. An essential function of a neighborhood is the way it performs the role of a safe place and acts as a temporary respite from the commuting lifestyles many of us maintain.

Physical features are probably the neighborhood's strongest defining character. An Australian report, 'Neighborhood Character', said it was building groupings, streets, public spaces, particular natural and artificial landscapes, and places such as community areas, shopping, and civic facilities which give a specific neighborhood its particular physical identity and for most people its meaning (Anonymous 1998). The report concluded that neighborhoods should contain well defined edges; fall within a comfortable five minute walking distance; have a clear hierarchy of streets and include a well-connected network for pedestrians, bicycles and cars; community facilities should have a neighborhood focus and develop local identity; and a diversity of housing types.

Wolf Von Eckardt's neighborhood has, "One common denominator...the basic necessities of life ought to be within easy walking distance. It is difficult to relate to others and feel neighborly while driving an automobile" (Von Eckardt 1978). Necessities include food stores, an elementary school and playground, neighborhood meeting place, public swimming pool, laundromats, dry cleaners, a tavern or two, and a park.

Finally, the authors of 'Planning to Stay' describe a neighborhood as "the basic social unit and physical building block of our cities" (Morrish and Brown 2000). Their neighborhood ideal was based on five key pieces of place—home and garden, community streets, neighborhood niches (markets), anchoring institutions, and public gardens (parks). These are broken down in terms of location, scale, mix, time, and movement.

(iii) Jane Jacobs Hyper-Local Perspective

Few have spoken more passionately about neighborhoods than Jane Jacobs. She celebrated neighborhood prominence in 'The Death and Life of Great American Cities' (Jacobs 1961). She challenged sociologist Reginald Isaacs's idea that the neighborhood in big cities may have no meaning. Isaacs argued that city people are mobile. They pick and choose from the entire city (and beyond) for their associations. Responding to his argument that neighborhoods were too provincial, Jacobs wrote,

> "Even the most urbane citizen does care about the atmosphere of the street and district where he lives, no matter how much choice he has of pursuits outside it; and the common run of city people do depend greatly on their neighborhoods for the kind of everyday lives they lead" (Jacobs 1961).

She identified two distinct kinds of city neighborhoods. The most intimate is the one we are familiar with at the local street level. This is the immediate space outside our front door in which we encounter people with whom we enjoy various degrees of casual relationships and from whom we can retreat into the privacy of our own residence. These are people we feel comfortable saying good morning to, or borrowing a proverbial cup of milk from, and who sometimes call on us or whom we assist in time of distress.

These neighborhoods work best when they provide many eyes watching the street throughout the day. These eyes enhance the protection of the innocent, such as children and the elderly, as well as sanctioning and supporting civic virtue.

The next appropriate level of neighborhood scale within the large city, she said, is the district. It may have 100,000 or more people. It not only provides for a variety of overlapping uses and has some form of political representation, but it is also an identifiable fragment of the larger city.

Districts are important because they are something to which many people belong, express their allegiance, and through which they organize resources in support of smaller street neighborhoods. Left alone, the latter would too often experience the futility of fighting losing battles because of their lack of political power.

Jacobs acknowledged a middle level of neighborhood as found in planning literature, the neighborhood unit. It generally has five to seven thousand residents, a number required to support an elementary school. She dismisses this configuration. "Why an elementary school?" she asked, when so many may attend a religious or a private institution. The Neighborhood Unit, as it is defined, is too large to support casual interaction but too small to allow for the diversity of activity and commerce of the neighborhood district.

Jacobs' neighborhoods are based on human vitality; listening to what she called the 'foot people'; connecting streets; diversity; organized complexity; the casual enforcement of safety; self-appointed public characters; mingling and mixtures of work and commerce with residences; the childhood charm of roaming up and down sidewalks; liveliness and variety; localized self-management; differences being superior to duplications; old buildings mixing with more recent; densities allowing for diversity; and even occasionally strange persons or places welcomed for the freshness of perspective they can bring.

The enemy of this living organism, she says, is the great blight of dullness. It erodes interest, caring, involvement, economic growth and experimentation. It puts in the same place only those who are similar in age. It has only one style of building. It generally tolerates only one use. It is too often thought of as a work of art, which must somehow be beautified. It is a place with problems requiring only a single prescription for their salvation. For Jane Jacobs a successful neighborhood,

> "Is a place that keeps sufficiently abreast of its problems so that it is not destroyed by them. An unsuccessful neighborhood is a place overwhelmed by its defects and problems and is progressively more helpless before them" (Jacobs 1961).

(c) Neighborhood Types

Obviously neighborhoods vary, but for our purposes the constant feature is that these are places in which people live. They may have a variety of additional uses or have none beyond households. They can be of varied density and height, range from publicly open to gated and closed, but their common element is human relationship of a generally casual, local, or implied nature.

Examining the form and content of these neighborhoods reveals three prevalent types, of which the most intimate is the 'street neighborhood', followed by, despite Jacobs' criticism, the 'neighborhood unit', and finally the 'neighborhood district'.

Like Russian nesting dolls in which the biggest contains the second-biggest, the second-biggest contains the third, and so on, the three neighborhood types fit into each other in a dynamic and living framework. The street neighborhood, for instance, is part of a larger neighborhood unit, of which there may be many 'roving' neighborhoods allowing citizens to pick and choose from the different service functions of distinct neighborhood units. These units are elements of an identifiable district neighborhood, which acts, as well, as a destination for outsiders.

It is the street neighborhood level at which most people experience the day to day sameness of their living place or on the other hand its improvement or decline. Uncovering those deeply sensed feelings of satisfaction merits attention by city planners and those engaged in the day to day process of city management.

(i) The Street Neighborhood (Fig. 2.1)

The street neighborhood describes those who live nearest to one's residence. It numbers anywhere from the few who live next door to us to the those on one's immediate street or perhaps those just around the corner. It could number up to about 1,200 or 1,500 people (though a number closer to 500 seems more likely) in the immediate streets but is dependent on area densities. It is limited to a street and its block surroundings or those who share several apartment floors or who live in a small building.

It is small enough that persons can maintain casual and usually recognizable encounters with others. It is at a level at which people are most comfortable providing assistance to each other. It recognizes that decisions can generally be reached on common issues. It is based on co-operation in advocating a particular interest for an area, is walkable, knowable, and provides the opportunity for many eyes to be on the street. Finally, essential services may be nearby, though in single use low density subdivisions this is unlikely to be the case.

The street neighborhood is our closest urban approximation to the village of several 100 members. It is usually part of a broader built area. It is not bound by kinship or a shared economy. At its most intimate level its population

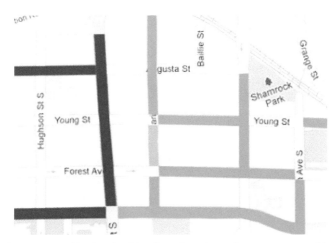

Figure 2.1. This street neighborhood in Hamilton, Ontario conforms to an area seldom exceeding the two to three blocks around one's home.

Source: http://renewhamilton.ca/wp-content/uploads/downloads/2014/03/HAMILTON-BRAND-Renew-Hamilton-Training-Program.pdf developed by the Regeneration Institute for the Great Lakes (ReIGL).

corresponds to the extended tribal family. Jane Jacobs, in deploring the dull and ambiguous urban places emerging in post-war America, described street neighborhoods as similar to early primitive settlements.

Christopher Alexander, a specialist in pattern language, said neighborhoods should not be "more than 300 yards (radius 450 feet) across, with no more than 400 or 500 inhabitants" [300 yards is approximately 275 meters, while 450 feet is about 137 meters]. Such a neighborhood would be from five to 15 acres [two to six hectares] based on either a square or circular configuration. This size was based on anthropological evidence that, "a human group cannot co-ordinate itself to reach [agreement on basic decisions about public services, community land, and so forth] ...if its population is above 1500, and many set the figure as low as 500." He referenced a 1952 report of the Philadelphia Housing Authority indicating that 500 is a realistic figure for organizing community meetings at the local level. As to physical size, Alexander cited a 1964 City of Philadelphia study reporting that people when asked which area they really knew usually limited themselves to the small area seldom exceeding the two to three blocks around their own home.

Small street neighborhoods of fewer than 1,000 people function within a larger urban structure. Size is limited by what might be described as an easy walking distance. A 40 acre (16 hectare) size might be its limit though it could be less in higher density areas and slightly more where conditions supported intimate neighborliness. The smallest unit of a Traditional Neighborhood Development (TND) as found in the New Urbanism approach of Duany/Plater-Zyberk is 40 acres, the upper limit of the street/block model. The Florida resort community of Seaside designed by the same team was created on an 80-acre village for 2,000 residents.

Village Homes in Davis, California, designed as a sustainable community, is a 60-acre, 242-unit mixed-use residential garden village. The St. Lawrence community in Toronto on 44 acres was designed as a tight knit part of the surrounding street pattern. Its diverse functions and an eight story maximum height ensured its seamless integration into the surrounding urban fabric with a density and population closer to that of a neighborhood district. Likewise the rebuilt Regent Park community in Toronto is being re-integrated into the surrounding urban fabric by re-connecting its streets to the city grid and providing for a range of housing types and a diversity of income levels amongst its residents.

(ii) The Neighborhood Unit (Fig. 2.2)

The neighborhood unit was a deliberately sized planning structure. With the development of large cities in the past 200 years, the neighborhood unit size was a key building block of residential plans. It was generally in the range of 3,000 to 10,000 people but its ideal size was around 5,000 to 6,000. It was on a

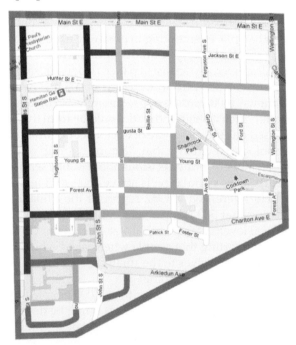

Figure 2.2. This Hamilton, Ontario neighborhood of Corktown corresponds to the neighborhood unit size.

Source: http://renewhamilton.ca/wp-content/uploads/downloads/2014/03/HAMILTON-BRAND-Renew-Hamilton-Training-Program.pdf developed by the Regeneration Institute for the Great Lakes (ReIGL).

land mass generally about 160 to 200 acres (64 to 80 hectares) though physical size might ultimately be defined by geographic or other artificial boundaries and so in some cases might be larger or smaller than these limits.

It was usually too small for political representation, particularly as we have moved towards greater political centralization as small municipal structures disappear. It corresponded to five digit zip codes, English wards, and Canadian census tracts. It was capable of supporting small retail and other non-residential uses though increasingly these have been rigidly separated from residences in new developments. Its primary role was to support one elementary school.

This neighborhood type has had tremendous resiliency. From the era of Greek city-states and even earlier, settlements have been complete entities of about 800 meters diameter, or a 400 meter distance from their center, within a 64 hectare space (160 acres). As urban places grew beyond that size and distance, a new settlement-type environment of similar size and function would grow alongside the old one. Dr. Matthew Hardy of the International Network for Traditional Building, Architecture and Urbanism (INTBAU), notes, "The boundaries of neighborhoods are not usually physically defined, but their centers—typically comprising a parish church, town hall, market, school and other public buildings—are usually found to be around 800 meters apart. This seems a relatively constant factor of human urbanism, defined by the limits of the body itself" (Hardy 2004).[1]

The optimal size of a walkable neighborhood today according to Dan Burden of the Local Government Commission Center for Livable Communities, is a quarter to a third of a mile from the outer edge to the center (400 meters or slightly more), or about a five to 10-minute walk at an easy pace. Burden's neighborhoods range from 40 to 125 acres [16 to 50 hectares], though a one third mile distance could be up to 225 acres [91 hectares].

Urban planner Clarence Stein, among others, used the neighborhood unit in his plans between the First and Second World Wars. New town planners Allen and Unwin, in 'The Size and Social Structure of Towns' released in 1943, recommended neighborhood units of 2,000 residences with between 6,000 to 7,000 persons and a range of communal facilities including places of worship, nurseries, primary schools, a library, child welfare clinics, playing fields, a public garden, a communal wash-house, laundry service, swimming pool, shops, restaurants, a cinema and a community center.

The unit's central place as a foundation of subdivision design was contained in Chiara and Koppelman's 'Urban Planning and Design Criteria',

---

[1] "This is to be a feature of all human urbanism, defined by the distance people are prepared to walk on a regular basis, about 400 meters or a five minute walk. Traditional cities tend to develop as a series of neighborhoods or 'urban quarters' of around 800 meters, in diameter, or about 64 hectares. Studies of traditional cities around the world demonstrate the ubiquity of this pattern." From Hardy, M., 'Renaissance of the traditional city', retrieved on-line 4/5/2004 from www.axess.se.

a reference source for the physical aspects of urban design and planning. Its model was a 160-acre (64 hectares) development with an elementary school to which no child had to walk more than a half-mile. Major streets did not pass through the development but were its boundaries.

For urban theorists Solow and Copperman, the ideal neighborhood supported an elementary school and required a population ranging from 2,000 to 8,000. They suggested a desirable size of about 5,000 persons. Depending on its density such neighborhoods might have to be up to 500 acres (over 200 hectares) with a half-mile or more walking distance between essential features. If a quarter mile was used as a more reasonable walking distance then unit size might shrink to 160 acres but even here, with so many twists and turns on increasingly disconnected and winding streets, there was no guarantee the quarter mile standard could be met. Their normal-sized neighborhood, sensitive to limits of publicly accepted densities, ranged from 50 to 250 acres.

Such sizes, however, can determine the ability of urban places to either evolve with time in response to different conditions or to wither and die as failed models of human living. Geographic layout of a neighborhood and its associated population either limits or expands the public's ability to experience a full array of urban features within walking distance, as well as supporting or restricting opportunities for civic involvement. They drive numbers necessary for effective public transit, nearby schools, and a multitude of retail necessities.

It should be remembered that while people reside in one street-level neighborhood they can feel part of and use multiple neighborhood units in a fashion described as 'roving' neighborhoods (Riemer 1950). While the above ideas may have their critics they do form the thinking of many contemporary proponents of sustainable communities (Roseland 1998).

(iii) The District Neighborhood (Fig. 2.3)

The district neighborhood is beyond a generally walkable size. A majority of its residents are strangers to each other. They share the sense of living in a well-defined geographic area with the reputation of being a destination, and one with several citywide and regional amenities. It generally has over 10,000 people, as many as 100,000 inhabitants and possibly even more. It is a collection of residential, commercial and other uses. It depends either on ambitious walking, public transit, or private automobiles, is of sufficient size to merit political representation, can act as an advocate on behalf of weaker street neighborhoods, is a known and generally popular destination for outsiders such as tourists, and has a distinct geographic identity based on a name, its history, and its positive or negative associations.

Until recently districts were the normal size of most big cities. Even Rome, which may have reached a million at its Imperial height, slipped back to the district size by the middle ages. Districts are the least precise in their size. Some can function as successful diverse areas well below the 100,000 figure because of accidents of location, natural boundaries, racial distinction, historic

Figure 2.3. Four downtown Hamilton, Ontario neighborhoods make up a neighborhood district. *Source*: http://renewhamilton.ca/wp-content/uploads/downloads/2014/03/HAMILTON-BRAND-Renew-Hamilton-Training-Program.pdf developed by the Regeneration Institute for the Great Lakes (ReIGL).

economic success, density, tourist appeal or the imagination and initiative of their residents. Others gain district identity for a combination of negative reasons such as diminished public safety, economic failure, or because they are essentially ghettos for a trapped or consigned population.

Districts match the historic size of the formative stages of many of today's major cities, but today they generally conform to what we would think of as a small city. A major city today is a combination of neighborhood districts, and major cities increasingly nest within larger city regions.

Popular and media discussion about neighborhoods often refer unwittingly only to district neighborhoods. They are branded with a title and may even appear on maps, street signs, and other general descriptions of city areas. They are often a destination for outsiders attracted by their ambience, character, and diverse uses. As well they can describe the social, physical and economic features of an area, track the value of its property, and contribute to the larger discussion as to which way a place's long term investment potential is tending.

As noted above Jane Jacobs argued that only large neighborhood districts could marshal sufficient resources for political action and economic diversity. They usually combine residences with a town center arrangement, including office and employment uses, night-time facilities such as theatres, museums, and hotel and retail such as major stores, speciality shops and restaurants. District neighborhoods of 100,000 are large enough to support a diverse economy and conform to politically representative areas.

(d) The Neighborhood as an Agent of Regeneration

Consideration now settles on the role of neighborhoods as settings for urban regeneration in which there are opportunities for both public and private sector partners. These opportunities proceed from key ideas about neighborhood identity and purpose, including:

- Walkability/Mobility
- Lifetime Quality
- Environmental Opportunity
- Their Economic and Job Creating Role
- Development of appropriate governance structures for comfortable and engaged existence
- Enhancing the bio-capacity to eco/carbon footprint ratio

While the latter four are the prescriptive pieces of an evolving neighborhood, it is the performance role of walkability, along with alternate mobility opportunities for the walking-impaired, which defines the comfort, safety, pleasure, necessary 'eyes on the street' and freedom of a place. It provides for ground level observation of local challenges, place-specific opportunities, and means of accelerating the move towards a neighborhood's 'lifetime' character. This latter character is the means not only of reducing the eco/carbon footprint but for building a bio-capacity enhancement process.

(i) Walkability

The local place has a future in an increasingly megapolitan world for the way it balances the outward flow to work, school, retail, etc. with an inward-looking perspective supporting creative unpredictability, serendipitous opportunity, economic plenty and enhanced public health. This 'intentional localism' is about the desirability of place and the free choice people make to live within a walkable area defined spatially by the 400 meter building block of urban design, but beyond which they have the freedom and ability to venture as the spirit moves.

So entrenched however is the historic record with tales of people searching for new economic and social opportunity, escaping oppression, and sometimes simply seeking the joy of the open road that one wonders how cities, towns, and villages could ever have developed, much less remained the keystones of most lives for good and ill (Saunders 2010, Rae 2003). Those who moved however have often been the exceptions while the rule remains that until fairly recently, the horizon of living for most people was the distance they might comfortably walk from home, and from which they might safely return, in a day.

Why would this have been the case?

Consider a hunting and gathering tribe of about 100 early humans having set up in an area, for discussion purposes, of approximately 16 square kilometers[2] or four by four kilometers. It would not have been square, but a rag-tag geography subject to the peculiarities of terrain and tribal needs (however for arguments sake consider the square to be the general shape). Their foraging would have been subject to limitations imposed by distances, a need for mutual protection (for those left behind as the hunters foraged), possible cultural factors,[3] and the bio-diversity of animal, plant life, and water available to them. Sooner or later however they would have exhausted, or polluted this resource base and so been compelled to move on to an equivalent sized new territory (Bouton 2011).

Only when they discovered the advantages of an agricultural economy was it possible to set up shop in the same area in a comparable space of about 16 square kilometers, and stay there almost indefinitely. As a result the merits of cleaning up their mess would become more urgent.

While our ancestors would not have had the measurement tools with which we are familiar, their common sense and native talent for determining

---

[2] There is an acknowledged arbitrariness to this measure and a tribe's territory may best be described as a series of ranges within which hunters traversed and so might have been considerably larger than this formulation. It is worth noting however, in a Guest Blog to 'Scientific American' by Tim De Chant on 16 August 2011, a tendency, as tribes grow in size, towards greater spatial density of occupation. A chart produced by anthropologists Marcus Hamilton, Bruce Milne and Robert Walker and ecologist Jim Brown shows 10 hunter-gatherers occupying a range of 5.6 square kilometers, 50 occupying a range of 10.8 square kilometers, and 100 occupying a range of 31.6 square kilometers. While such ranges would actually contradict the case for an increasing occupant spatial density per square kilometer, the exponent of 0.75 they use, would, if applied to the territory occupied by a tribe growing from 50 to 100 members, round out closer to 19 square kilometers. Each hunter-gatherer tribe has different territory sizes as the researchers conclude, "Some hunt more, some gather more." *Source*: http://blogs.scientificamerican.com/guest-blog/2011/08/16/hunter-gatherers-show-human-populations-are-hardwired-for-density/.

[3] C3. In her review of 'How to Think Like a Neandertal' by Wynn and Coolidge (Oxford University Press 2011), Katherine Bouton says of Neanderthals, that "…they were xenophobic, occupying a small territory from which they rarely strayed." In referencing Wynn and Coolidge's book she notes the likelihood that "…between one percent and four percent of the genome of non-African humans is derived from Neanderthals. And distant matings may have played a role in human culture."

time and space, based on seasonal variations, position of the sun, and condition of the land would have provided them with different means for interpreting their world.

For us 16 square kilometers is composed of foundational measurements for which we have an intellectual as well as a physical sense. Four kilometers is close to the old imperial measure of 2.5 miles, while a kilometer consists of 1000 meters or about 3/5ths of a mile. As a general rule of thumb from time immemorial most people can cover the former distance, at a reasonable walking speed, in approximately an hour.[4] So a 2000 meter or 1.25 mile walk out from a settlement's either temporary or permanent location within the center of a 16 square kilometer territory, and a 2000 meter, or 1.25 mile return walk, would have been consistent with limitations described above.

No wonder then, that from the time in which hunter-gatherers stalked their survival needs, to the agricultural villager's outward bound passage from the diminishing chimes of the village's church bell, or similar religious soundscape, this was a generally universal daily limitation of travel and boundary of their known territory.

Breaking down the 4000 meter (4 kilometer), diameter of one's familiar territory, and its walking threshold, into smaller more intimate urban measures makes it easier to understood the prevalence of the 400 meter distance as a signature piece of urban design, which can be walked in about five or six minutes (one's pace depending on how far one is going).

The 400 meter distance is at best an average but one with powerful meaning and is a functional correlate of most things we consider building blocks of good neighborhood design from the near-by-ness of schools and public transit, the limiting distance people are prepared to walk for necessities, to the space within which children are permitted to wander, and even as a measure of personal health (Pollowy 1973).

Modes of getting about in North America occurred initially within a grid formula for road layout and property distinction. It was more formal than the European structure which had evolved over centuries in a less rational fashion. The old parameters of distance however with their time-honored roots were naturally transferred to a New World still dependent on walking, the cornerstone of the 400 meter building block.

The bicycle was a briefly populist means of travel available by the late 19th century to a broad based distribution of the population. It fit well within the grid system. It broke down the old restrictions imposed on men and women who could neither afford a horse, or other means of transportation to escape their village life, but it was quickly surpassed by the automobile whose nature and use rejected considerations of walkability and conflicting human presence.

---

[4] For the purposes of this discussion I am assuming about five minutes to walk 400 meters if that was the limit of one's walk, but that in a walk of 4000 meters the person's pace would adjust to the longer distance by moving at a rate of six minutes for each of the component 400 meters.

As the crowded, mean streets, of inner city life were abandoned for the dead-end cul de sacs of the new suburbs they appeared to be an able replacement for the dangers, real and imagined, of the traditional city. They limited automotive access but also restricted the ease of pedestrian movement from one place to another. More perversely, in removing the clutter of work, commerce, and other apparently conflicting uses from these residential neighborhoods, they ensured the car's eventual ascendancy.

About the only remnant in these new suburbs of the old 400 meter measure of acceptable walking distance was a local school despite its being compromised by today's multiplicity of school types. Some schools retain their neighborhood function but overall there has been a decline of children walking or biking to school from 66 to 13% in the period from 1974 to 2000 (Brody 2012). Parental concerns for their children's safety, given an absence of eyes on the street to watch out for them, as well as the preponderance of cars dropping off children at the school, has made such destinations less safe for those who walk.

The 400 meter distance may even have deep seated meaning for life—literally! A study in the 'Journal of the American Geriatrics Society' indicated amongst those who predicted their inability to complete a 400 meter walk there was a 91% likelihood (with a sensitivity of 46% and a specificity of 97%) that they were in fact unable to walk 400 meters (Sayers et al. 2004). A report from the University of Pittsburgh published in the 'Journal of the American Medical Association', indicated moreover that the inability to complete this 400 meter walk was a significant predictor of death and poor health in the elderly (Newman et al. 2006).

Robert Roy Britt, LiveScience Managing Editor was blunt in his conclusion, 'Walk a Quarter-Mile [400 meters] or Die', and that for men and women aged 70 to 79, "If you can walk a quarter-mile [400 meters], odds are you have at least six years of life left in you….and the faster you can do it, the longer you might live" (Britt 2006). Failure to do so, other studies suggest, limits one's life span to three years or less.

If the 400 meter distance is a means for measuring one's longevity, and its completion by a relatively healthy adult can be undertaken in five minutes (though somewhat more time would be required by children and the marginally infirm), it is a damning indictment that we have and continue to build places with so little diversity, not only within that range, but far beyond it. Walking in such places can be a dreary experience of sameness and generally pointless given the absence of anything other than residences of varying densities.

According to Garry Peterson, a 'walkable urbanity', has certain requirements.

> "…the enticement to walk is key to making density wonderful, to providing realistic transit options, to making smaller greener homes compelling….The big thing I think falls out of most walkability

formulas is a quality critical to the actual experience of walkability, and that's the extent to which the place in which you live is connected (by walking routes and easy transit) to other places worth walking to.... The true test of walkability I think is this: can you spend a pleasant half hour walking or on transit and end up at a variety of great places? The quality of having a feast of options available when you walk out your front door is what I am starting to think of as 'deep walkability'" (Peterson 2010).

Dom Nozzi expanded this idea to all the advantages of walkable urbanity including health, the serendipitous experience of meeting others in the neighborhood (what Dan Burden describes as the benefits of 'Bump Into's Per Square Minute'—BIPSM), as well as reductions in noise, air pollution, and impermeable surfaces (Nozzi 2009). Decrease in the latter would have a notable knock-on effect of mitigating heat island problems, and reducing the infiltration of excessive storm water into sewers and which, in its untreated form, often ends up in the very water bodies from which drinking water is drawn. "Walkability", he says "exists when there is convenient access. The home is so close to a park, a grocery store, a movie theatre, places of work, nightlife and civic institutions that it is an easy, short walk to nearly all of life's daily destinations. Car ownership must be optional if a walkable lifestyle is to exist."

An additional performance feature is 'human scale', the ways in which the built and natural environment meshes with our personal sense of safety, comfort and joy. Kirkpatrick Sale[5] has described the relational size of buildings relative to the street (front yard, sidewalk, verge and roadway) and whether the humans walking within this setting experience either personal satisfaction, or feel so uncomfortable, that given a free choice, they would not return (Sale 1980).

Less easy to define are aspects of an intentional localism such as sustainability—a place's resilience in the face of unexpected climatic or economic eruptions far removed from local control; a place's conviviality—

---

[5] Sale argues. "... the most successful residential streets, those that satisfy in some indefinite way as you walk or drive along them, are those with houses set back from the street 50 feet or so; farther than that and the buildings tend to get lost in the background, nearer than that and they give a sense of crowding, of looming" (Sale 172), Burden, D. 1999. Street Design Guidelines for Healthy Neighborhoods. Local Government Commission: Sacramento. According to Burden the concept of the enclosed street or outdoor room derives from the time of the Greek Empire. "... traditional street designers have achieved this comfortable sense of enclosure by giving streets a ratio of 2:1 to 3:1 width (from building to building) to building height. ... People walking along the street like to feel that they can "reach out and talk to someone" sitting on the front porch, which is possible when porches are within 20 feet of the sidewalk" (p. 29).

it's hard to define charm, occasional mystery, and it's amazing distinction;[6] a place's ability to grow old with architectural dignity—encouraging users to support its regeneration; and finally its accommodation of a variety of unexpected, unusual, and necessary features once deemed unacceptable by local zoning by-laws but now encouraged within a broader performance criteria (Lynch 1960).

(ii) Lifetime Quality

The other significant performance feature is the concept of the 'Lifetime Neighborhood'. A discussion paper released by the Department for Communities and Local Government in the United Kingdom reflected a public interest in addressing the needs of an aging population. The underlying notion however, not explicitly stated but implied, is that one could freely choose, if other aspects of life supported such a choice, to reside for their entire life in one place (Dept. for Communities and Local Government 2007).

What would such a place look like? Firstly it would respect the generally accepted 300 meter radius young children are permitted to roam. Many eyes on the street would keep them within public view, roadways would be designed to allow them to safely navigate street corners, and parents could be confident in extending this small act of independence to their children (Jacobs 1961).

For teenagers it means there are places to hang out with friends and things to do, as well as opportunities for small jobs, and age-specific entertainment.

For parents it means that a family feels secure in this location, that nearby housing is affordable, and that there is access to medical, schooling and other support systems. And as aging takes place there is the opportunity to downsize from one's residence and have access to appropriately designed living places while remaining in the neighborhood, along with an increased reliance on nearby support services, and neighbors and family supporting one's comfort.

Living places would be of multiple sizes, and purposes, so that leasing options, small houses, one storey residences, monster or heritage homes, group homes, apartments and a range of other residential possibilities would be available to accommodate one's needs at any stage of life. Public transit availability within the appropriate 400 meter building block structure, while

---

[6] Lynch explored the importance and changing quality of cities in terms of what is seen, remembered and delighted in; Jackson, J.B. 1980. The Necessity for Ruins. University of Massachusetts Press: Amherst., said "The search for sensory experiences of the world as the most reliable source of self-knowledge is more insistent than ever" (17); exploring these layers of meaning was an exhibition, 'Sense of the City', mounted by The Canadian Centre for Architecture (26 October 2005–10 September 2006), "dedicated to the theme of urban phenomena and perceptions which have traditionally been ignored, repressed, or maligned." The subtlety of this discussion was described in a special issue of the British football magazine, 'When Saturday Comes', titled 'At the Match Special', WSC May 2005, p. 4, "… there are five aspects of the experience of going to the match, less tangible but important nonetheless, that have been lost: *visual purity, physical density, individuality, unpredictability* and *anonymity*."

respecting the limitations of public resources, would be as robust as the density of residence permitted. Likewise schools, retail, medical support, and workplaces, would be permitted, even encouraged, uses as part of a performance obligation.

Ultimately however it means neighborhoods, scaled to the sizes described above, should have the features of conviviality, splendor, variety, joy, and all the other attributes of a happy and engaging place. If all neighborhoods conformed to this quality we would have resolved many of our environmental, economic and civic challenges.

(iii) The Environmental Opportunity

Neighborhood features have environmental consequences. The use of the neighborhood idea for instance as a founding principle of modern urban planning influenced, for better or worse, suburban places built in the immediate post war era. More recently, the planning philosophy of the Congress for the New Urbanism, along with a rating system for neighborhoods, developed through the LEED (Leadership in Energy and Environmental Design) accreditation process, reflect a greater attention to the environmental impact of neighborhood design.

The neighborhood is at the service of environmental opportunity. It is a key observation of proponents of the Congress of the New Urbanism that settling the issue of human habitability may be the most important piece for renewing environmental health (Duany 2000). The argument can be summarized as follows—to the extent we find satisfaction in our living places, confront the continuing privatization in our lives, measure the matters that are meaningful to healthy families and friendships, and limit our dependence on cars, global resources (with their impact on local places far away) and excessive consumerism, we will tackle environmental quality.

The ways of addressing these issues have often been disconnected from daily lives except for small measures like weekly recycling. In combination with civic participation and economic inventiveness however they provide for a confrontation of ideas in which cost savings from greater energy efficiency are not an excuse for larger homes or commuting longer distances to work. Designing locally for human scale provides for personal comfort, a sense of shared intimacy, a place for meaningful encounters, and the ability to reinforce and expand social capital.

Emerging forms of technological infrastructure however may provide the greatest opportunity to break from a past of highly centralized and dumb infrastructure. Hybrid and green options, provide increasing opportunities for neighborhood-based renewal at the street, unit and district level.

Cascading levels of local wastewater treatment for specific uses, and street and household level storm water management, are moving past the experimental stage of application. Renewable demand-driven energy initiatives supported by natural gas and even mini nuclear options are on the horizon. A circular waste regime rather than a cradle to grave formula could operate at the largest district scale of neighborhood.

These distributed options are combined with the inter-operability embedded in smart IT supported infrastructure (Moffatt 2001). They provide local work opportunities, the chance to enliven and broaden the kinds of uses within walking distance, and a new set of facilities potentially managed by some form of neighborhood governance. These measures are applicable in both developed and developing countries and allow for decentralization, democratization and distribution of resources downward to those using them.

Such places broaden public choice beyond those which are aesthetically dull, environmentally harmful, dangerous to health, and lack the kind of spirited animation of historically successful and vibrant places.

(iv) Locally-based Economies

Associated with developments such as those above is the promise of a new kind of locally-based, artisanship economy, of which its fully realized form can only be somewhat glimpsed. Businesses ranging from small-scale 3-D printed manufactured designs, to local production and distribution from food processing to beer-making, are such features (Davidson 2012). So too is a move away from centralized, far away infrastructure solutions to a local delivery/treatment of waste, water and energy, with opportunities not only for local work but neighborhood entrepreneurism (Moffatt 2001). Just as new platforms of information technology and digital communication are already revolutionizing many aspects of business, they are starting to change how we think about and provide all kinds of services from medical to schooling, and energy to water.

Economically, the integration of global economies, the increasing concentration of sectors like retail among fewer alternatives, and the sprawl inducing features of logistics management, create their own counter reaction of underground economies and community-based activities. These may have short existences or struggle to survive but they continue to appear and occasionally flourish. They respond to our market interest in diversity, new ideas, and individual initiative.

Third places in local communities, such as bars, coffee shops, and even the basements of places of worship are one such economic opportunity (Oldenburg 1989). They are the informal meeting places beyond home or work. Despite the prevalence of home electronic entertainment, or the homogenizing aspect of regional entertainment and shopping, they meet a human need for familiar encounter, provide convivial spots for dining and drinking, and are places for sharing information and access to new services and products. Too often however they are banned by local by-laws.

(v) Civic Dimension

The civic dimension of a move towards a 400 meter building block of urban design would be the opportunity to re-envision how local services are provided and the ways appropriately sized levels of local government could operate within an urban dominated world.

All of this permits a different kind of environmental response to the continuing increase in sinks and a decrease in sources (Owen 2010),[7] founded on replenishing bio-diversity by lowering and even reversing impact at its source (in urban areas) and allowing the external countryside from forests to meadows, and woodlots to wetlands, to perform their eco-system service role more effectively.

What prevents any of this? Often it's a sense of entitlement that in an 'Oliver Twist-way', pleads, *"Please, sir,* I want *more"*, only in this case the demand is driven by profligacy not poverty; a belief that we have no obligation to meet the needs of others so 'Not in My Backyard' is a reasonable response to change; and a preference for the soul-destroying character of single-use living environments in which there are few reasons to walk, and from which, as people age, they will have to leave just as they are becoming comfortable with the place.

At the civic level the changing character of urban governance, as in some ways more important than the nation state and in virtually all cases more significant than the state/provincial level, finds an emerging need for regional/megapolitan governing structures to handle issues of mobility, energy use, and water quality protection. Such structures however are even further removed from the everyday understanding and connection of citizens. The neighborhood in all three of its forms provides an opportunity for experimenting with different styles of more intimate and local authority, while generating a broader obligation beyond an exclusionary exclusivity.

(vi) Bio-capacity to Eco/Carbon Ratio

Why is the neighborhood important to the bio-capacity to Eco/Carbon footprint ratio?

As more activities and services are delivered at the neighborhood level, the carbon impact of car use is reduced. More significantly however while car use for other megapolitan purposes remains (though possibly within a new

---

[7] The rebound effect in which energy and environmental enhancements cause increased environmental harm is described by Owen (2010). "The problem with efficiency gains is that we inevitably reinvest them in additional consumption. Paving roads reduces rolling friction, thereby boosting miles per gallon, but it also makes distant destinations seem closer, thereby enabling people to live in sprawling, energy-gobbling subdivisions far from where they work and shop." Refrigeration, satellite radio, and general energy consumption fall into the same conundrum of modern life.

ownership paradigm), there is an increasing attentiveness to the quality of these neighborhood places. Almost without exception these tend towards acts which are of a regenerative quality—investing in well-known and loved spaces and buildings, adding to bio-diversity through tree planting and gardening measures, supporting initiatives to allow people to age in place. It is infused as well with the opportunity for locally based environmental solutions, economic opportunities and civic enhancement.

The chart below provides a starting point for thinking about, measuring, and implementing regeneration solutions scaled to the appropriate neighborhood size. Before examining it however a few interpretive comments are required.

Roving units listed below are those 'unit-sized' places in close proximity to each other and which are shared in their use by the householder, though the latter's particular 'street-level' neighborhood will only be part of one of these units. These roving units and potentially other units with no connection to one's household, or regular lifestyle, make up a neighborhood district.

Across the top of the chart below are the varieties of neighborhood types as described above. Along the left hand side are areas to be addressed within a regeneration imperative.

While it's clearly difficult to calculate the two fold increase in bio-capacity alongside any initiative's one fold eco/carbon footprint impact, nevertheless this becomes an exercise worth refining over time. One does so by interpreting the value of green initiatives listed under bio-capacity increase as against either the eco/carbon footprint impact of producing, delivering, installing and maintaining such initiatives, or, as another considered factor, the diminishing eco/carbon footprint resulting from alternate mobility means beyond car use, of the shortening of transmission distances for necessary, just-in-time and IT monitored services, or the benefits associated with on-site management of water, waste and energy.

Civic describes the ways of engaging people and developing new modes of governance suitable for varying neighborhood levels.

Economic opportunities describe those entrepreneurial possibilities emerging out of a neighborhood level regeneration, from those for the local handyperson, or teenager seeking small jobs, to agents versed in installation processes, maintenance and emerging operational alternatives (Fig. 2.4).

(e) Renewing the Neighborhood Role

There are at least four possible outcomes for renewed and regenerated neighborhoods. The first is market driven as privatizing communication technologies create opportunities for local connection, or as new products (locally grown food) and services (home delivery and neighborhood retail) make this imperative the preferred choice. The second is imposed, as environmental degradation and depleted energy sources force market transformation and government action, either by targeted taxation or strategic

Examples	*Household/ Street*	*Unit*	*Roving Units*	*District*
Bio-capacity increase	Rain barrels; green roofs; walking and mobility options	Green verges; tree planting; bicycle connections	Parks and woodlots as appropriate	Wetlands; water treatment ponds; composting centers
Eco/carbon Impact	Acquiring produced products	Installation; delivery of items	maintenance	Further measures to reduce car use
Civic Participation & Governance	Informal/ formal participation	Neighborhood council ->	Co-operating councils	Formal governance
Economy-business and Jobs	Handy-person opportunity; jobs for young people	Local services; third places ->	Work co-ops and regional businesses	City-wide Employment teams and businesses

Figure 2.4 Opportunities Emerging at Different Levels of Neighborhood Regeneration.

use of resources for best end uses, thus forcing people to work and shop closer to home. In the third, fear or antipathy to outsiders, perceived threats from terrorists or petty crime, and self-interest, promote covenanted communities in which homeowner associations and deed restrictions impose obligations from paint color to hours of swimming pool operation.

Lastly however is the intentional route in which attention is directed to matters of civic participation, economic control, and environmental damage causing a deliberate shift to a neighborhood-based sense of responsibility and obligation. The opportunity is profound; the verdict is unclear.

References

Anonymous. 1998. Neighborhood Character: An Urban Design Approach for Identifying Neighbourhood Character. Australia Department of Urban Affairs and Planning. New South Wales.

Bouton, K. 28 December 2011. If Cave Men Told Jokes, Would Humans Laugh? New York Times.

Bradford, N. 2004. Place Matters and Multi-Level Governance: Perspectives on a New Urban Policy Paradigm. McGill Institute for the Study of Canada Annual Conference. Challenging Cities in Canada, 11–13 February 2004.

Britt, R.R. 02 May 2006. Walk a Quarter-Mile or Die. LiveScience Managing Editor, www. livescience.com.

Brody, J. 30 January 2012. Communities Learn the Good Life Can Be a Killer. New York Times.

Calthorpe, P. and W. Fulton. 2001. The Regional City. Island Press 31.

Chaskin, R.J. 1995. Defining Neighborhood: History, Theory and Practice. Chapin Hall Center for Children. University of Chicago.

Courchene, T.J. 2004. Citistates and the State of Cities: Political-Economy and Fiscal-Federalism Dimensions. Retrieved August 5 2015 from www.ppm-ppm.ca/sotfs/courchene.pdf.

Department for Communities and Local Government. November 2007. Towards Lifetime Neighborhoods: Designing Sustainable Communities for All, from www.communities. gov.uk.

Duany, A., E. Plater-Zyberk and J. Speck. 2000. Suburban Nation. North Point Press, New York, p. 151.

Epstein, H. 12 October 2003. Enough to Make You Sick. New York Times Magazine.

Florida, R. 2002. The Rise of the Creative Class. Basic Books, New York.

Hardy, M. Renaissance of the Traditional City, retrieved online 4/5/2004 from www.axess.se.

Hayden, D. 2003. Building Suburbia: Green Fields and Urban Growth. Pantheon Books, New York.

Jacobs, J. 1961. The Death and Life of Great American Cities. Random House, New York.

Katz, B. 2004. Neighborhoods of Choice and Connection: The Evolution of American Neighborhood Policy and What It Means for the United Kingdom Metropolitan Policy Program. The Brookings Institution Research Brief.

Keller, S. 1968. The Urban Neighborhood. Random House, New York.

Lynch, K. 1960. The Image of the City. M.I.T. Press, Cambridge.

Moffatt, S. 2001. A Guide to Green Infrastructure for Canadian Municipalities, FCM Centre for Sustainable Community Development.

Morrish, W. and C. Brown. 2000. Planning to Stay: Learning to See the Physical Features of your Neighborhood. Milkweed Editions, Minneapolis.

Newman A.B., E.M. Simonsick, B.L. Naydeck, R.M. Boudreau, S.B. Kritchevsky, M.C. Nevitt, M. Pahor, S. Satterfield, J.S. Brach, S.A. Studenski and T.B. Harris. 2006. Association of long-distance corridor walk performance with mortality, cardiovascular disease, mobility limitation, and disability. JAMA 295.

Nozzi, D. Measuring Walkable Urbanity, retrieved 22 September 2009, www.walkablestreets.com.

Oldenburg, R. 1989. The Great Good Place. Marlowe and Company, New York.

Owen, D. 20 December 2010. The Efficiency Dilemma. New Yorker Magazine.

Peterson, G. 11 January 2010. Deep Walkability for Sustainable Cities. Retrieved 14 January 2010, http://rs.resalliance.org/2010/01/11/deep-walkability.

Pollowy, A.M. 1973. Children in the Residential Setting. Universite de Montreal, Centre de Recherches et d'Innovation Urbaines: Montreal.

Putnam, R. 2000. Bowling Alone: The Collapse and Revival of American Community. Simon and Schuster, New York.

Rae, D. 2000. City: Urbanism and Its End. Yale University Press, New Haven.

Riemer, S. 1950. Hidden dimensions of neighborhood planning. Land Economics 26(2): 197–201.

Roseland, M. 1998. Toward Sustainable Communities Solutions for Citizens and Their Governments. New Society Publishers, Gabriola Island, BC.

Sale, K. 1980. Human Scale. Coward, McCann & Geoghegan, New York.

Saunders, D. 2010. Arrival City. Knopf, Toronto.

Sayers, S.P., J.S. Brach, A.B. Newman, T.C. Heeren, J.M. Guralnik and R.A. Fielding. 2004. Use of self-report to predict ability to walk 400 meters in mobility-limited older adults. Journal of the American Geriatrics Society 52(12).

Scanlan, T. 1984. Neighbourhood Geography. Is Five Press, Toronto.

Solow, A. and A. Copperman. 1948. Planning the neighborhood. Chicago. Public Administration Service.

United Way of Greater Toronto and the Canadian Council on Social Development. 2001. United Way—Poverty by Postal Code: The Geography of Neighborhood Poverty—1981-2001.

Von Eckhardt, W. 1978. Back to the Drawing Board: Planning for Livable Cities. Simon & Schuster, New York.

von Hoffman, A. 1995. Local Attachments: The Making of an American Urban Neighborhood. Johns Hopkins University Press. Baltimore; xix.

Example of a Green Building

Royal Botanical Gardens Atrium (CGBC 2012)

The Royal Botanical Gardens (RBG) is a designated Natural Historic Site in Canada and is situated among 1,100 hectares of exceptional natural gardens and spectacular sanctuaries. The Atrium addition (named the Camilla and Peter Dalglish Atrium), features a number of sustainable and innovative initiatives and provides a balance of plant and human comfort.

Location: Burlington, Ontario
Size: 1,487 m² or 16,000 ft²
Certification: LEED Canada Gold NC

Green Technologies and Features

Ventilation

- A biofilter living wall is featured and creates a natural air purification system. Rainwater harvested from the roof provides irrigation needs and supplies water closets.
- The three-story height in this facility is conditioned by variable air volume displacement system that is supplemented by an air-side heat recovery and hydronic in-slab heating. Thermal comfort is maintained through controlled temperature stratification.
- Temperature and humidity levels are carefully controlled all year round to balance the environmental needs of the vegetation and the visitors.
- Fully retractable glazed walls and the associated glazed roof function to control ventilation and temperature requirements as well as providing an extraordinary experience for visitors to interact with nature and take in panoramic views of the botanical gardens and the Niagara Escarpment.

Energy

- The building utilizes an advanced lighting control system that includes occupancy and daylight sensors and multi-level switching and dimming. Exterior lighting is controlled by photocells.
- Motorized and fixed shading devices actively and passively manage the solar heating.
- A reduction in energy consumption of 63% compared to comparable facilities.

Water

- Rainwater harvesting has provided an opportunity to reduce the demand on the municipal water usage by 40%.

Picture and Description, Courtesy: Tony Cupido

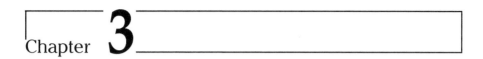

Chapter **3**

Practicing Regeneration

Thinking about our Challenges: A Big Picture Contemplation

The City of Toronto (bound by Steeles Avenue in the north, Lake Ontario in the south, Etobicoke Creek and Highway 427 in the west and the Rouge River in the east) encompasses approximately 630 square kilometers, or 243 square miles, of territory within which over 2.6 million people live.

It is a land mass with both high density apartments and low density bungalow-type dwellings. It has parklands, water surfaces, river valleys inhabited by wildlife and feral cats, lands used for transportation, schools, businesses, and commerce. A visitor would agree however that it provides most amenities required for a good life as well as sufficient open spaces. One does not, with a few notable exceptions, feel crammed into a 'Hong Kong-type' setting.

Now imagine that all seven billion 'plus' people alive today lived in an area with an overall population to territory distribution comparable to Toronto's. Then imagine us reaching 10 billion at some point in the century.

How much of the Earth's surface, or perhaps how many Canadas, would be required to house those populations?

Turns out it's not even all of Canada, in fact it's not even close. Using the 'Toronto area population to territory ratio' the seven billion would fit in a space comparable to less than 1/5th of Canada or about the size of the northern Canadian jurisdiction of Nunavut. The 10 billion would require about a quarter of Canada.

This exercise, admittedly a fantasy, does show that as a species we can and would be able to, given our tendency towards increased city living, require much less space on which to live. In some ways we are already moving towards this land use.

1. Trending Towards Regeneration

Several years ago, a Doug Saunders' article in the Toronto Globe and Mail, 'The Hush-Hush Regreening of Europe', reported a curious but seemingly underreported phenomenon occurring throughout Europe (Saunders 2007). Millions of hectares of land formerly used for marginal agricultural purposes were being turned back into forest land more suited to the soils and geographic conditions of their discreet place.

Nor were these forests bleak wastelands of trees but places generating their own economic opportunities for recreation, food stuffs, cleaner water, renewed passages for a suddenly more robust nature, and not incidentally acting as carbon sinks. The latter consideration is now informing the thinking about what to do with marginal farmlands in places like Central America,[1] particularly if carbon trading or taxes ever become internationally implemented.

This is a foundational character of contemporary regeneration as described for those lands outside of megapolitan regions or smaller urban centers, and places described as countryside, rural, wilderness, and even the middle landscape (Marx 1964).

The practice Saunders describes is not new. In many cases forests throughout New England, New York State and in eastern Canada are 20th century creations within a landscape in which the original old growth forests had been stripped clear for farming in the previous century. Erosion along with other environmental maladies such as reduced ground source water, and the marginal utility of these lands for farming had caused a re-thinking and then a re-planting of the forest, such that what is recent looks as if it had been there from time immemorial.

New economies and opportunities emerge different from those dependent on farming and in the process result in dislocation and adjustment. Such upheavals characterize much that passes for regeneration. We have enough experience now to know they are often a necessary by product of changes required either because of the failure of what has been on the ground or because of the need to envision new responses to persistent challenges.

On a global scale we are experiencing something similar driven by changing demographics as more people flock to cities but also by an increasing awareness of the threatened character of a diminished bio-capacity and the necessity of not simply acknowledging the significance of natural capital but aggressively supporting measures to add to its prominence (Sustainable Prosperity 2014).

While many environmental prescriptions have been based on lowering standards of living with the consequent object being a reduction in our ecological and carbon footprint, such occurrences are unlikely to be the result

---

[1] http://www.face-thefuture.com/en/projects/ecuadorsierra describing a community *afforestation* project in the Ecuadorean Andes.

of well-grounded policy but will arise some suspect from our overuse of both natural resources and those other services provided by nature resulting in a punishing and sad retreat into something recognizable as a modern 'dark age' (Jacobs 2004).

More optimistically and embedded in the promise of regeneration however is the sense that many factors line up well with a more bountiful and optimistic future. This despite the likelihood that dislocation will occur and those pragmatic environmental choices will have their own problematic aspects.

Successful regeneration is an approach that moves beyond prescription and towards performance thus it is tailored to the peculiarities of place, is flexible in considering appropriate options, and ultimately is driven by an outcome in which bio-capacity enhancement exceeds eco-carbon footprint impact, and in the process not only enhances broader social and international economic prospects but provides for a creative and still emerging labor intensive opportunity.

A 2:1 heuristic ratio of bio-capacity renewal as against an increase in the eco/carbon footprint, has been suggested herein as a tentative target, though there are obviously short and long term means of calculating this guideline.

Elements of this thinking are beginning to permeate public policy in spirit if not directly in application. In Norway for instance the country's, "Sustainable Development Strategy requires that decreases in the country's natural capital (such as through extraction of oil and gas) must be offset by increases in other forms of wealth" (Sustainable Prosperity 2014).

New York City's Green Infrastructure Plan released in 2010 put a dollar value on the financial benefits accruing from every fully vegetated acre of green infrastructure in terms of its reduced energy demand, reduced CO_2 emissions, improved air quality and increases in property value.

> "The plan estimated that a combination of green infrastructure, cost-effective grey infrastructure and other program elements … will result in a net reduction in combined sewer overflows of 40 percent by 2030 while saving US $2.4 billion through costly investments in traditional infrastructure such as tanks and tunnels" (Green Infrastructure Ontario Coalition and Ecojustice 2012).

This 'bio-capacity to eco/carbon footprint' big picture thinking is reflected in Saunders description of the reforestation processes underway in Europe. "A hectare of forest", he writes, "Will 'sink' 150 tonnes of carbon, easing the continent's [Europe's] Kyoto Protocol commitments on reductions of greenhouse gas emissions" (Saunders 2007). Saunders cites a 1990s initiative accounting for 1.5 million reforested hectares of land, and another ending in 2006 that might have added an additional million hectares, with a combined price tag of US$5 billion.

It's quite possible, he says, that equal amounts of unrecorded land-use changes are being generated by farmers abandoning marginal lands in the wet, rocky and mountainous areas largely in southern Europe and the Balkans. Without both these deliberate programs and fortuitous acts it's quite likely that measurements of carbon in the atmosphere would have been even higher with increased environmental impact.

As noted in earlier chapters the flocking together of human populations in diverse and opportunity-laden cities is a good thing. An emptying-out countryside which of course will never be fully abandoned allows for bio-capacity enhancement on a massive scale, and that's a good thing as well. Alongside this is an emerging recognition that 'lean' can actually be more fulfilling and healthier than fat, and associated with this the knowledge that information technology while destroying many one-time flourishing occupations also opens the door to inter-connected solutions at a local level facilitating new kinds of work, civic engagement, and robust environmental measures suited to their performance in a discreet place rather than through centralized prescription.

This is neither wishful, nor economically detrimental, thinking. A variety of industries depend on predictable environmental conditions from tourism, forestry, and ocean fisheries to the insurance industry, winter sports and casual recreation, as well as municipal services and transportation schedules.

All of us depend on ecological services from clean air and water, to the carbon sequestering capability of trees and the storm water holding capacity of wetlands. We are gradually placing a valuation on such services as a means of targeting not only their protection but their participation in the economy itself.

Operational and lifestyle profligacy has allowed us to be wasteful of resources with extraordinary impediments to personal and corporate bottom lines. Environmental destruction is simply bad for business and is tolerated only because of an affluent abundance in the developed world. "Carbon in the supply chain is not only a liability in itself but an indicator of an inefficient operating model and exposure to rising energy prices," says a productivity consultant at Ernst and Young (Harvey 2008).

Public health challenges and their impact on the public purse are at least partly dependent on how urban design limits physical activity, while the impact of air pollution, car accidents, and food quality, to name a few, are all aspects of environmental conditions.

While no one can be certain of the real economic opportunity encompassing regeneration, the reported potential expense for new and rebuilt infrastructure worldwide will be in the order of US$40 trillion (Booz Allen Hamilton 2007). A significant portion of this investment, together with measures for reclaiming or rethinking how these projects are undertaken and managed, will rely on partnerships with the processes of regeneration.

Regeneration is based on rebuilding much that has been lost or damaged, and in the process creating a refreshed economic discussion in which growth is aligned with renewal. At its core is an active engagement process which incorporates re-building and re-connecting the quality of natural places supporting bio-diversity; adding to the stock of resilient ecosystem services including those supporting carbon sequestration, climate change resiliency, clean air and water; renewing and revitalizing existing human created assets from buildings to safe, diverse, convivial and vibrant communities; and anticipating, even in new projects, the need for their eventual restoration and re-purposing (i.e., mining, and housing design) (Cunningham 2002).

Arguments for and against adaptation of the built and natural environment often collide with proponents for more aggressive mitigation strategies, but as Joyce Rosenthal argues in issue 37 of the Harvard Design Magazine, "…these arguments frequently fail to recognize the carbon-reducing co-benefits of many adaptive strategies in the built environment; techniques such as urban forestry and green and reflective roofs cool the urban environment while also reducing energy consumption. For many planning and design strategies, carbon mitigation and climate adaptation are synergistic and mutually reinforcing approaches" (Rosenthal 2014).

Urban resilience, she argues, has supplanted sustainable development as the focus of urban planning and design given the necessity of a city being able to absorb shocks and disruptions and remain functioning. Perhaps ironically such measures often cost less to build, contribute to predictable outcomes for many industries, and generally look after themselves without expensive human and mechanical input and updating.

It also takes the discussion of the continuing quality of built and natural places, regions, and the biosphere beyond a self-deluding notion that incremental improvements, small disconnected pieces of conservation, and one-off 'greening' initiatives will significantly reduce the damaging direction of our current biases towards not only resource depletion, and massive reduction in global bio-diversity, but also character-less sprawl. Regarding the latter, a Congress for New Urbanism gathering in Buffalo in June 2014 was told there are 10,000 square miles in the United States Great Lakes watershed alone zoned for single family housing with potentially calamitous effect on near-shore community waterways flowing into the Great Lakes. The cumulative damage of these biases has the power to destroy even individual private realms which once might have thought they were beyond public impact.

Accordingly we now proceed to a consideration of a regeneration approach in three areas of the built and natural world. The first are the wide expanses of open space beyond the megapolitan fringe which as noted above we describe as countryside, middle landscape, rural abodes, and wilderness. The second are the individual buildings and infrastructure in the built environment regardless of location. The third is within the megapolitan region and the broader reasons and opportunities for tackling the possibility of retrofitting single use, low density car dependent places. All three, but particularly the third require tackling the nature of appropriate governance for such a challenge.

2. Countryside Bio-capacity Building

(a) The New York City Potable Water Model and Its Roots

Up to 90% of New York City's drinking water is provided by the Catskill/ Delaware system, covering 1,600 miles. It is the largest unfiltered water supply in the United States. It is estimated that watershed protection programs, including outreach and education, land management, land acquisition, and partnerships with watershed non-profits and municipalities costs about US$100 million a year. It begs the obvious question—wouldn't a combination of filtration and disinfection be a simpler way to achieve healthy standards?

The answer is a resounding no! Even an effective filtration and disinfection program would be insufficient, concludes the Environmental Protection Agency, without preemptive watershed protections. Construction cost alone for a filtration plant large enough to support the Catskill/Delaware system would be in the neighborhood of up to US$10 billion dollars, with annual operation and maintenance costs comparable to what is being spent on watershed protection.

> "Without preemptive land protections", the Prince William Conservation Alliance concludes,[2] "Significantly higher levels of disinfecting chemicals would need to be used to purify the water. By-products from these chemicals are known to pose serious health risks and their use is in fact limited by federal law. Finally, high levels of chlorine are believed to damage the fittings within the water delivery system. For all these reasons, it is preferable to protect the drinking water at its source."

We take water for granted. Such inattentiveness has a cost in over-engineered and expensive centralized infrastructure. As the above illustrates however an alliance with natural systems can provide for investment significantly below that required for physical structures.

It explains why programs to encourage re-planting of forests and other natural systems are extolled by regeneration proponents for bio-capacity enhancement.

New York State was one of the first to inaugurate such a strategy. Protection of lands in the Adirondack and Catskill Forests dates back to 1885. Explicit regeneration has its roots in the early 20th century with the establishment by the Forest, Fish and Game Commission of a tree plantation in the Catskills in 1901 to replace what had been lost through aggressive lumbering.

Mark Kuhlberg's historic examination of the gradual transition from the clear cut harvesting by both New World settlers and the lumbering industry to a more intentional understanding of the value of forested lands, documents a kind of inevitable evolution in thinking, akin to where we find ourselves today.

---

[2] http://www.pwconserve.org/issues/watersheds/newyorkcity/.

"Communities soon began experiencing the unanticipated consequences of this environmental degradation. In many of the watersheds which had been denuded of their forest cover, water tables dropped precipitously, rendering uncertain formerly dependable water supplies. Stream and river levels fluctuated dramatically as the frequency of destructive flash floods increased. Erosion became a serious problem, as topsoil which no longer retained its moisture was blown away. In extreme cases, exposed sandy soils were swept into neighboring fields of crops and across roads, damaging viable farmland and creating as much of an impediment to travel" (Kuhlberg 1974).

There were few more enlightened proponents of regeneration measures in the United States than Franklin Delano Roosevelt (FDR) who between 1912 and 1943 oversaw the planting of over a half million trees on his family's thousand-acre Hudson Valley estate. As first a member of the New York State Senate and later Governor he followed in the political footsteps of his fifth cousin Theodore Roosevelt in using the mechanics of governance to expand on what had been a personal understanding and initiative.

FDR's Civilian Conservation Corps established on his becoming President in 1933 was a direct reaction to the catastrophic impact of the Great Depression. It was an act of foresight and ambition to prevent soil erosion, control floods, and eventually the planting of 221 million trees. Many of the measures of this period have been supplanted over time as we learn more about the process of regeneration. Conifer plantings were a quick way to return forest to the land by stabilizing erosion, improving the watershed, and gradually restoring depleted soil with their eventual detritus. These monoculture regimes are in decline and are more susceptible to attack by invasive pests but they have fulfilled their purpose and can now be replaced by a more diverse selection of native hardwood and softwood species as natural succession enhances the woodland's overall health.

(b) The Ontario Reforestation Model

In Ontario, Edmund Zavitz joined the provincial Forestry Bureau in 1903. With a BA from McMaster and a Masters in the Science of Forestry from Michigan State, he was a legendary advocate for the value of forests, and a virtual Johnny Appleseed with the creation of a forest tree nursery and the distribution of free seedlings.

His landmark 1908 'Report on the Reforestation of Wastelands in Southern Ontario' identified two classes of land for permanent management as forest. Kuhlberg writes.

"The most evident were the 'large, contiguous areas of non-agricultural soil.' These he identified as two sand plains, one along the north shore of Lake Erie (in Norfolk County) and the other in Simcoe County, as well as the sandy moraine which ran laterally through the counties of York, Ontario, Durham, and Northumberland….Secondly, Zavitiz noted that there were isolated patches of poor soil throughout what was otherwise good farmland" (Kuhlberg 1974).

Much of what Zavitz imagined and recommended has come to pass. Ontario's current greenbelt encompasses much of the sandy moraine he recommended for management. Tree growing plantations were likewise established in three provincial locations.

George Linton ran the provincial nursery at Orono, Ontario, 60 miles northeast of Toronto in the 1920s. He undertook extensive silvicultural experiments to determine best practices for soil reclamation, irrigation, and fertilizing techniques, and in time this work contributed to regeneration projects after the Second World War of which the most noted is the 11,000 acre Ganaraska Forest, the largest block of continuous forest in southern Ontario.

Regeneration of those areas outside the major metropolitan areas thus has a long and distinguished history, yet in many ways it might seem like the hard work was done a long time ago and many might question its relevance to today's challenges. In the early 1990s two Ontario governments of widely different ideologies terminated support for private tree planting programs throughout the province and only recently have these been reinvigorated. Today it is more likely to be intensive cash crop farming that threatens woodlots than urban expansion (Fig. 3.1).

It is the heroes, legacy and model constructed by long ago regeneration pioneers which inspires today's bio-capacity performance strategy.

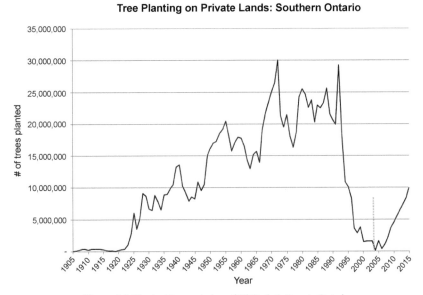

Figure 3.1. Used with permission of Al Corlett, Forests Ontario.

(c) The Buffalo Commons and the Great Plains

One such discussion surrounds what has been called the Buffalo Commons, a project to return almost 140,000 square miles of land in the dry portion of the American Great Plains to its somewhat original native prairie and in so doing reintroduce the American bison into parts of Montana, Wyoming, Colorado, Oklahoma, New Mexico, Texas, North Dakota, South Dakota, Nebraska and Kansas. Originally proposed by Frank and Deborah Popper in 1987 it was in response to periodic calamities such as the Dust Bowl and the continuing de-population of an area unsuited to small-scale farming.

Native grassland and a more restricted ranching use would constitute the Commons. The economies of First Nations and remaining inhabitants would revolve around a culture closer to that first encountered by European settlement. The region's major water resource, the Ogallala Aquifer would be less stressed; tourists might flock to an area totally unique in character; and the carbon sequestration potential of bio-diverse native prairie land would be realized.

A critical perspective however was provided by Joel Kotkin in his 2012 Great Plains study for Texas Tech University. He argued that the territory should not be abandoned for this less intensive use but continue to provide opportunities for high tech and large scale agriculture, natural gas and other drilling activity, pipeline infrastructure, renewable energies including wind, biofuels and ethanol production from large scale corn and related growing regimes,[3] as well as manufacturing, research and development with an information age connection.

While acknowledging a continuing rural depopulation he cites increasing growth of urban regions and, though he doesn't reference the megapolitan character of major urban centers in commuting proximity of each other, this does appear to be the emerging reality for at least some parts of the contiguous territory from Dallas Forth Worth through Oklahoma City and Tulsa and, stretching up to Kansas City and Omaha. A quick look at the map of this area however shows that most of the remaining territory (with the exception of Denver), is and will continue to be an open, slowly depopulating landscape, as even Kotkin recognizes:

> As cities such as Oklahoma City, Sioux Falls, Midland, Lubbock and Omaha grow, smaller communities close to these hubs will also expand, while those further away from roads, airports, and natural attractions will continue to diminish (Kotkin 2012).

---

[3] Mother Jones magazine reports 'King Corn Mowed Down 2 Million Acres of Grassland in 5 Years Flat' between 2006 and 2011, from http://www.motherjones.com/tom-philpott/2013/02/king-corn-gobbles-climate-stabilizing-grassland-midwest. It begs a significant policy issue as to whether grassland, a long-time (though one itself subject to human intervention) and in-place ecosystem, should be replaced by a water demanding, genetically modified, pesticide regime for corn-ethanol and soya production servicing cars, future hamburger meat, and industrial uses.

In other words, like the apparent conflict between Kotkin's suburban reality as against that of Duany's new urbanism, it appears both futures might be realized but in ways reshaping the character of the other. There's little reason the Buffalo Commons concept, for at least a significant portion of these lands, and the urban opportunity of its major centers, can't coalesce in a boarder vision. The Buffalo Commons concept is simply following the over 100 year lead of places like New York State and the abandoned countryside of part of New England.

(d) Resolutions

So why do we have such problems addressing these challenges? Far too often our differences of opinion are based on narrow ideological perspectives or prescriptive self-interest rather than pragmatic and performance approaches. Where and how might a fresh perspective be mounted?

First would be the necessity of identifying those lands with impermeable coverage below the 10% cut-off for damage to watersheds and managing them as permanent sites for carbon sequestration, water management, wildlife enhancement, and bio-capacity enhancement. The 'Development Status of Forested Watersheds' in Canada prepared by the World Resources Institute and Global Forest Watch Canada,[4] provide a sense of where this process might begin.

Land in the 'over 50% developed' status includes a broad swath of Atlantic Canada through Nova Scotia and New Brunswick, along the St. Lawrence Coast through Quebec, then covering most of southern and near northern Ontario, proceeding as a line through urbanized Manitoba and Saskatchewan before taking up much of Alberta (The Oil Sands region and the urbanized portions of the province) and eastern British Columbia as well as the urbanized area around Vancouver. Contiguous areas show development of between 10 and 49%, but major regions of Labrador, northern Quebec and Ontario as well as a band stretching from northern Manitoba and Saskatchewan through the territorial jurisdictions of Canada are below the 10% figure. On a national scale such regions should be maintained and enhanced in perpetuity.

One then begins the process of expanding the map to include those places capable of immediate remediation to drive them below the 10% impermeable factor. Finally one not only directs urban development to areas significantly above the 10% figure but also addresses either the greening of impermeable surfaced places through locally specific on-site solutions as described in the sections below, as well as maintaining and enhancing the continuing bio-capacity role for areas within them in which hard surfaced coverage is significantly less than the area average.

---

[4] http://www.globalforestwatch.ca/files/publications/20010200A_gfwc_xrds_report.pdf.

3. Net Positive Development Ethos and What It Means

In championing the cause of more sustainable suburbs, the new dean of Princeton University's School of Architecture, Alejandro Zaera-Polo, bemoaned the car dependency of his community, despite a tendency of people in Princeton to walk or cycle when possible, and said, "We have an obligation to build more sustainably through smarter material choices and energy systems" (O'Connor 2014). Radiant floors, double-glazed glass, and tilt-and-turn window systems as part of a natural ventilation system were some of the items he recommended.

There is however a necessity for even more substantial measures in suburban neighborhoods if such places are to raise their sustainability profile. In their re-engagement with nature the most profound changes will either be implemented intentionally or by necessity as unpredictable weather, to paraphrase an old saw, 'waits for no person'.

(a) Living with and within Water

Few responses better characterize the shallowest interpretation of ways forward than the debate over what to do in the greater New York region in responding to the likelihood of a repeat of a Hurricane Sandy-like catastrophe, which flooded the homes of those near the Jersey Shore and even shut down the New York subway system.

Cold water has been thrown on the theory that hurricanes like Sandy are necessarily or solely caused by global warming. Even the latter idea of climate change is awash in critics downplaying its significance. Less debatable are the cold (perhaps a better description would be hot) facts of ocean levels rising due to a combination of warming seas and melting glaciers. This trend will continue (Englander 2012).

The Dutch, for whom most of their country is below sea level, long ago recognized the crippling power of water on everything from utilities to homes. Along with a series of national measures such as dykes, they have introduced more local means such as water-capturing and water avoiding construction.

In the New York region this might mean a macro system of barriers and pumps, turning parks into water-containment barriers, creating a greenbelt of parkland around urban areas, and micro solutions such as green roofs to soak up rainwater, along with trees and permeable pavement to enhance groundwater infiltration. Or it might include none of the above but simply rebuilding the storm walls damaged by Sandy in the Far Rockaways.

Dutch architect Henk Ovink was hired as a senior adviser to Shaun Donovan, the US secretary of Housing and Urban Development, to undertake a more resilient approach for the built and natural environment in the event of future catastrophe. He quizzed the Far Rockaways engineers as to whether they were building the same storm walls that had broken. When they said yes, he asked what they would do if the walls gave way again. "We'll rebuild them again", he was told (Shorto 2014).

Of course it's hard to abandon centuries of thought about nature, that it's an enemy that must be tamed, encased or fenced off. The Dutch solution has been to live with it and within it.

The power of regeneration is the way it changes the intellectual and financial equation. Brent Dorsey of Entergy Corporation, a utility on the US Gulf Coast said his company can't move its business to an off-shore location in another country. If they do nothing however their facilities and operations might literally move off-shore into the surrounding water bodies. There is a strong need therefore for adaptive measures to deal with a changing climate and future climatic events.

Resilience, beach nourishment, wetland restoration, barrier island restoration, elevated homes and green roof and wall type measures are part of a suite of strategies for embedding protective measures against future disaster.[5] They have implications for that company's bottom line. Calculated as an aspect for avoiding catastrophic loss their benefit is measured at a 1:4 ratio of 'cost to install' versus the anticipated loss and the cost of rebuilding if no action is taken.

(b) Net Positive Development

Few have spoken more passionately about the necessity of not only designing with, but embedding nature itself, in the built world from infrastructure to individual buildings, than Janis Birkeland. Her ideas are summarized below.

"Nature", she writes, "Has provided for the infrastructure and basic services to support human life, and has even subsidized our profligate Western lifestyles. Now, however, we have exceeded the earth's carrying capacity. We have also exhausted the cultural and social viability of many resource-rich, but impoverished countries and colonies around the world... genuine sustainability will require more than social change and ecological 'restoration'. It will require increasing the total amount of ecosystem goods and services, as well as increasing the health and resilience of the natural environment" (Birkeland 2009).

Birkeland promotes an eco-innovation approach in which natural systems and environments replace unnecessary machines or products. Accordingly any development should increase the size, health and resilience of natural systems. Infrastructure should by definition regenerate, so that it flourishes and delivers ecosystem goods and services in perpetuity.

From a practical point of view what does this mean? In urban areas the horizontal and vertical space of individual buildings and multi-storey structures, by retrofitting their walls, roofs, and alleyways, would provide an

---

[5] From Building a Resilient Energy Gulf Coast: Executive Report http://www.entergy.com/content/our_community/environment/GulfCoastAdaptation/Building_a_Resilient_Gulf_Coast.pdf.

'edge' for habitats allowing them to generate eco-services. Vertical wetlands, breathing walls, multifunctional atriums and sunspaces, living machines, and solar ponds are just some of the means for undertaking this.

Why is eco-retrofitting not being realized to its potential? In part, she argues, it's because intellectual paradigms, institutional frameworks and power relationships, work against it. Our educational systems, she warns, have neglected design in favor of accounting skills. We seek templates and tools that tend to circumvent and subvert design so that even many 'green' design norms are rules of thumb that avoid innovation and can actually run counter to sustainability in certain sites and contexts.

She sees opportunity in existing suburbs. They have space to generate positive impacts, such as water and soil remediation for nearby land, or more extensive bio-diversity habitats on the site than existed before development. Yet she decries the failure to increase overall ecology in the suburbs.

Most green design has been at the building scale, Birkeland writes, but there has been less effort at greening whole neighborhoods. Eco-sanitation systems, like wetlands and living machines, are usually most practical at that scale. Merely increasing residential density with greener buildings could preclude land uses that provide eco-services beyond the needs of individual homes.

Our personal choices, she concludes, are often between a more efficient lawn mower or planting native grasses or even xeriscaping one's property so that no additional watering is required. They are choices between electricity for lighting and air conditioning or relying on passive solar measures and extensive tree planting.

Some of the biophysical performance features of Birkeland's net positive development include:

- Resources autonomy,
- Increased use of ecosystem services,
- Restoration and increase in natural habitats and bio-diversity,
- Reducing land coverage while increasing natural capital in buildings
- Designing and managing development to improve human and environmental health,
- Replacing resource and capital intensive machines with natural microbes
- Using organic, compostable, adaptable and renewable resources,
- Increasing natural and national security, avoiding fossil fuel dependency, decentralizing energy, and encouraging urban farming,
- Maximizing solar heating, cooling, and ventilating.

4. The Megapolitan Region and the Challenge for Governance

"Regions", Mallach and Brachman reported in 'Regenerating America's Legacy Cities', "Need their legacy cities' assets while the cities need the regions to fulfill their labor force needs and to better distribute the burdens of urban infrastructure and other costs. Commuting patterns, health care services, and employment bases all show increased interdependence among urban, suburban, exurban, and rural areas as people live in one suburb, commute to another suburb or to downtown, and shop in a third part of the region. Moreover, residents in more rural sections of the region often seek health care in the suburban or urban areas" (Mallach and Brachman 2013).

This supports earlier observations about people's existential lifestyles and psychological identities not being confined within a jurisdictionally distinct municipality but ranging across a megapolitan region. The construction of appropriate governance systems reflecting this reality awaits both political will and models for supplanting historically implemented but increasingly irrelevant local and state/provincial entities. The challenge alone for managing watersheds and human mobility within the megapolitan region already befuddle a system with jurisdictions either too small (municipal) or too large (state/province).

Since this is a megapolitan conundrum, at the larger level a successful regeneration strategy requires agencies with overlapping authority consistent with Mallach and Brachman's observation that the European experience, "… where a combination of regulatory controls and affirmative policies reduced sprawl and enabled cities to maintain their central roles even as they lost much of their industrial bases, illustrates that urban collapse and uncontrolled sprawl are not inevitable products of post-industrial economies, but flow from the interaction of those social and economic forces with destructive public policies" (Mallach and Brachman 2013).

For most people however this level will always appear somewhat abstract and subject to the wisdom and good guidance of a professional class of managers.

The neighborhood level as described earlier provides the essential territory for embedding opportunity, transparency, obligation, civic engagement, and democracy though one also requiring experimentation, imagination, and risk. It encompasses a series of scaled settings in which a regeneration imperative is authentically tackled. Its performance elements below also provide the essential features with which a renewed governance model would engage, and include:

- Appropriate intensification measures supporting a variety of living places for all ages and incomes (the British concept of Lifetime Neighborhoods),
- Design considerations of graduated density arrangements rather than harsh, side by side, low and high density configurations,
- Opportunities for citizens to hire and work with architects producing infill projects,
- Attention to Kirkpatrick Sale's notions of human scale, and, associated with this, human comfort and safety in the built environment,
- Encouragement of third places (Ray Oldenburg) such as restaurants, coffee shops, pubs, within walking distance,

- Public art opportunities, and places for creative expression and performance,
- Resting places and green spaces,
- Reconnected and narrower streets,
- Providing reasons and vistas supporting walking, along with design enhancements like naked streets, and woonerfs,
- Digitized infrastructure management cognizant of system interoperability, and integration with green and hybrid solutions,
- A metaphysics of place based on sounds, smells, and sights which create impressions unique to each recipient and often have qualities some describe as sacred, others as mythic, and others as artistic, and
- Distinct and distributed environment and energy projects.

(a) Getting There from Here

At the 2014 Canadian Urban Manufacturing Summit in Hamilton, a former Vaughan (municipality on the northwest edge of Toronto) planning and development commissioner, Frank Miele, presented research from York University's school of public policy on the top-five reasons why an industry chooses to locate in a municipality. Tax rate wasn't one of them!

"Number one: quality of life", said Miele "Quality of life is about education. It's about wonderful parks and a good transportation system, all these things that you take for granted" (LaRusic 2014).

Additional factors include a city's attractiveness, pride in the local community, and the ability of people to get to a company's workplace which for many means effective public transit.

If we regenerated our neighborhoods in support of this performance we might solve many of our most perplexing problems. These approaches not only restore but revitalize, re-urbanize, and intensify, and in the process provide associated environmental enhancement, civic transformation, economic development and improved public health. In this performance-based pursuit, while the greenest neighborhood or building is potentially the one already built, its regeneration is the most effective strategy.

One means of encouraging people to live in walkable mixed use locations would be lower property taxes for places meeting the above criteria. Low density car dependent places impose higher costs by the need for servicing a smaller population base (more fire halls to ensure nearby response; wider streets to maintain; water and sewer infrastructure stretching further into one-time countryside; and more police whose primary duty is traffic related because of the auto dependency of such places). Given the general affluence of persons increasingly living in walkable mixed use places however the inequity of them paying less in property taxes than those living in places of low density car dependency is problematic.

A better approach for the latter places might be the use of tax incentives for performance-based regeneration aimed at their gradual transition to walkable, mixed use and diverse places.

The ongoing conundrum however is that the very process of regeneration enhances the attractiveness of places, even those some distance from the old city core. In so doing it raises property values so that eventually these places are no longer affordable either for lower income families unable to pay the rent in what had been their long time neighborhood, or for a striving middle class. Hence these people are often forced to move even further outside the old city's limits and spread out in that vast though connected region we call the megapolitan place.

Solutions include a robust and connected mobility system allowing people to get around efficiently and by means beyond the automotive. Means for keeping people in place in older, suddenly attractive, neighborhoods might be an increase in geared-to-income rental opportunities. The financial cost of such policies may be worth it but the alarms of dread from comfortable and well off neighbors against further development (and particularly development of this type) may be prohibitive from a political and legal point of view.

(b) The Witton Lofts Project

To reinforce points already made, acts of urban regeneration are disruptive, but in the final consideration they are necessary if an area is to survive, flourish and tackle today's challenges. Individually however they are often unseen, or barely noticeable, even to the most attentive viewer. Some encounter unforeseen challenges not anticipated on first viewing, while others reveal possibilities unknown to previous occupiers of a space. Nothing better demonstrates these features than The Witton Lofts project on Murray Street north of Hamilton, Ontario's downtown just off James Street and within a short walking distance of the waterfront.

At ground level one wouldn't even know the waterfront was that close, though one would hear the shunting of trains at the nearby CN Rail yards. The view changed for Steve Kulakowsky of Core Urban Inc. (a company describing their mandate as creating architecturally relevant and interesting places for people to live, work and play) when he climbed to the top of what had once been the Mcilwraith School. Built about 80 years ago, it had been occupied for the last 25 years by the city's Mission Services.

Kulakowsky said, "I remember going up on the roof of the Mcilwraith building and realizing I could see the waterfront and the extension of Lake Ontario. Such a view however wasn't possible from a room on even the highest floor of the existing building. So we decided to re-develop the building as a condominium project, name it after its original architect, William Palmer Witton, and add three stories to provide a premier space for purchase."

The developers immediately faced three challenges that might have daunted less committed urban regenerators. First, was the attitude of the residential neighbors. How might they react not only to construction in their surroundings but the heightening of what was already a significant physical presence? The North End Neighbours are an association with a formidable and deserved reputation as advocates for their community.

Kulakowsky was adamant however, "If the neighborhood association had objected and wanted to fight us we'd have walked away from the project. There were just too many other challenges to make it worthwhile. But we talked to them, showed them our plans, told them how we planned to add three stories to the top of the existing structure, and that our intent was to sell the completed condominium units to an upscale market. They liked what we had to say, and perhaps were relieved that a different clientele would be occupying the building from its previous tenants' audience."

Challenge number two was the nearby CN (Canadian National) rail yards. Architect Rick Lintack of Lintack Architects described the uniqueness of the problem.

"CN Rail's proximity is such that they exercise development controls through an easement, effectively preventing significant residential change. Their switching yard generates a lot of noise which could bring complaints from neighbors. To forestall that they control what can and can't be done locally. In this case Steve Kulakowsky and his group negotiated with CN. In the past CN objected to operable windows that when opened allowed residents to hear rail activity. So we went with a solarium concept which blocked the sound from entering the inner part of the residence." It worked and CN gave the project the go-ahead.

Now came the biggest challenge. The old school building was physically sound in its own right but, says Kulakowsky,

> "Like any building that old it had settled over the years and though structurally sound the existing frame could not handle any additional stories nor could the existing concrete plate at its base accommodate such an addition. So we basically had to put a building on top of the existing building by threading 30 columns through the old building below and putting a sleeve around them so they wouldn't disturb the existing concrete plate. Anyone looking at the building of course only sees one structure but in fact it's like we put a building on stilts above the existing one."

In the case of Witton Lofts one could theoretically remove the old school building below and the upper three stories would remain on the pillars holding them up. Rick Lintack tells how it was done.

"Steve and his group wanted to convert the building and expand upwards. Originally two additional floors were envisioned but in our 3D modeling we felt it lent itself to a three story addition. We retained a structural engineering firm, MTE out of Burlington, and they concluded that the addition was feasible, but because the structure is concrete its ability to accommodate additions could not be determined. As a result a complete new structure was built on top of the existing one. It is supported by pillars using a helical pier system. There are new footings and the independent base was built by cutting into the existing plate."

Today natural daylighting streams through upper windows. Full occupancy has been achieved. A heritage structure has been repurposed. Intensification has been furthered in the downtown and the condominium is a short walk from a GO transit link to Toronto scheduled to be ready in 2015. For his part Steve Kulakowsky was recognized in 2014 with the Canadian Urban Institute's NextGEN Award for his commitment to downtown Hamilton's economic development and championing the cause of other young developers.

(c) Regeneration as a Cultural Imperative

Another and perhaps the most refreshing aspect of the regeneration imperative at work is the manner in which it has gradually assumed a kind of cultural imperative in which untold numbers of households reject the primacy of a grassed and desolate front and backyard lawn and a monoculture of similar planting.

Italian gardens in big city backyards have long been a wondrous demonstration of this unorthodox refusal to obey one type of vision and each is often a tangle of vines for small batches of home-made wines. Likewise tomatoes and herbs provide for a distinctly 'terroir' laden taste.

In the American mid-west thousands of individual households, a small token perhaps but a symbol of the adage that a free market works best when it represents the imagination and free choice of many decision-makers, are revolutionizing the idea of what one's small piece of residential property can and should be. They are restoring their yards to native prairie with its requirement of far less water, no fertilizer, and an ecosystem alive with wildflowers.[6]

For others it's ensuring the greatest amount of bio-diversity in their front and backyard, in some cases allowing weeds to flourish, and even planting milkweed to enhance the survival rate of Monarch Butterflies. In other cases it's installing mini-apiaries to render the life of the essential honey-bee less stressful. An irony of modern living places is the often greater health of urban bee colonies over their rural counterparts due to the higher and legal pesticide regimes allowed in agricultural settings.

Slowly an impetus for change is informing policy makers as well as public and private practitioners. Natural cover re-introduced on larger expanses of property supports improved run-off retention, erosion control, wildlife habitat reclamation, natural system connectivity, and improvements in air, water, and soil quality. The benefits of wetland cover, particularly at times

---

[6] 'In Midwest, Bringing Back Native Prairies Yard by Yard' from http://e360.yale.edu/feature/in_us_midwest_restoring_native_prairie_ecosystems_kessler/2603/.

of catastrophic storms, are finally being recognized for their water quality filtration, enhancement of wildlife and fish habitats, run-off retention and erosion control.

Permeability measures on once hardened landscape support groundwater infiltration, ecosystem health, and ground and surface water quality. Riparian cover, and even fish ladders, on creeks and rivers enhances water quality, fish habitat and diversity, systems connectivity and water temperature management. Daylighting of streams formerly lost in underground, human-constructed, tunnels restores an additional element of bio-diversity, and enhances nearby property values.

A resulting healthier water system not only supports ecological and human health, it protects human life and property, and finally improves the quality of those bodies of water into which it flows such as North America's Great Lakes perhaps the region's greatest resource for not only potable drinking water but fresh-water dependent industries.[7]

Alongside these are continuing experiments growing foods in vertical farms and in the open spaces of abandoned city blocks in places like Detroit; green roof installation and associated planting regimes; permaculture, community gardens and the simple add-on benefits accruing from on-site and backyard planting. In Paris for instance there are hundreds of small wineries dependent on even the tiniest spaces, sometimes recessed into residential courtyards, for propagating vines.

All of the above may meet only a small fraction of human food needs but their potential for further experimentation cannot be denied.

Application of bioswales, engineered wetlands, storm water ponds in dense urban settings, bio-retention planters, permeable pavers, boulevard and verge area bio-retention units and even parklands in a bowl configuration to allow for recreation most days but also to hold the rainwater of severe storms are aspects of a process for "reinventing storm water management", and ending the sole dependence on end of pipe solutions (Green Infrastructure Ontario 2012).

Some of these measures are criticized by those who note that in returning the landscape to prairie, well-intentioned practitioners are relying on a regime of toxic pesticides and controlled burning to prevent the natural succession of shrubs and trees—though burning at least was the very methodology practiced by the indigenous population long before the arrival of Europeans.

Critics argue it isn't natural, but honest patrons of the movement recognize the manner in which all actions are a kind of choice between competing poisons. The art of regeneration is disturbing and disruptive but in re-organizing both the built and natural world, a healthier and more resilient place can be created.

---

[7] Ceres water report on industries (computer chips, beer production, pharmaceuticals, apparel, agriculture, etc.) dependent on fresh water, and a Buffalo dean of architecture's prescription for eventual increased habitation of failing Great Lakes cities because of the abundance of fresh potable water now lacking in the American southwest.

5. Case Studies in Acts of Natural (Ecological) and Built (Urban) Regeneration

(b) Case Studies

Case 1: The Essence of Natural Regeneration: adapted from Guterl 2012. The clear lowland ponds in which Marten Scheffer, a biologist at Wageningen University in the Netherlands, had grown up swimming, had by the 1980s, turned turbid. Plants died, algae covered the surface, and only bottom-feeding fish remained. Fertilizer runoff from nearby farms was the culprit but even after it was stopped the ponds remained dark and scummy despite replanting them with lilies and stocking them with trout.

Mr. Scheffer's key insight as described by Guterl (2012) was, "The ponds behaved according to a branch of mathematics called 'dynamical systems', which deals with sudden changes. Once you reach a tipping point, it's very difficult to return things to how they used to be. It's easy to roll a boulder off a cliff, for instance, but much harder to roll it back. Once the ponds turned turbid, it wasn't enough to just replant and restock. You had to get them back to their original, clear state."

Scheffer removed the fish that thrive in the turbid water, recognizing that they stirred up sediment, which in turn blocked sunlight from plants, and ate the zooplankton that kept the water clear. "His program of fixing the Netherlands' ponds and lakes is legendary in ecology," Guterl summarized.

Case 2: The essence of built regeneration: adapted from Newman 2001. On a recent summer evening on a sidewalk in Brooklyn, two middle-class black men approaching 40, met on a Brooklyn sidewalk and noticing the out-of-state license plates of cars passing by got talking about the three R's: race, riches and real estate.

Newman picks up the story:

"Look at these cars", said Dezo El, pointing at the sedans and S.U.V.'s hugging the curb of South Elliott Place. "Vermont. My man here is from Connecticut. That car over there says Virginia. The woman across the street has North Carolina plates."

The other man nodded his agreement. "I don't mind when a community changes", said the man, a photographer with a salt-and-pepper beard who gave his name only as Gregory. "But the way it's changing, it's not changing for me and you. It's changing for them." He turned to a passer-by, expecting an amen. But the passer-by, Earl Avery, was having none of it.

"You can't say it's not changing for the better for us", Mr. Avery said. "A lot of black folks who owned homes here took the money and ran, moved to Staten Island or down South or wherever. I live here, and it's changing for the better for me."

Avery a former gym teacher and one-time resident of the Fort Greene projects was not only a homeowner but had purchased several buildings in the neighborhood. Gregory is a renter.

Gentrification in the Fort Greene district was raising the rents of those in brownstone apartments causing many to move out. DeKalb Avenue meanwhile was filling up with hip new restaurants.

Newman writes, "Some of the prime beneficiaries of the boom in Fort Greene (besides the real estate agencies that form an economic development zone unto themselves) have been middle- and working-class blacks: nurses and mailmen and civil servants who bought their houses cheap in the 60s, 70s and 80s and kept the community together while most other black neighborhoods in the city were falling apart."

(b) Observations and Conclusions

Sudden change characterizes both of the above 'communities' but in the case of the lowland ponds their restoration was made possible by an engineered solution carried out by ecologists, engineers and other construction and technical specialists operating within the natural environment. This kind of regeneration can have unintended negative consequences or fail to deliver the anticipated results. At the same time it may damage a well-loved and aesthetically pleasing, though invasive or alien, ecosystem.

Some might wince at the effect of these actions on the living non-human attributes of the pond including its current plant life and the fish that thrive in the turbid water, but we accept such surgery as the cost of restoring the ponds to a healthier, more robust, and resilient condition.

Urban regeneration does not have the luxury of such dispassionate engineering. In the case of the Brooklyn neighborhood, change is the result of market forces, human choices, and external factors. The ultimate destination of the neighborhood and its residents is uncertain. Objectors however cannot be as easily ignored as the fish which once thrived in turbid waters. Deliberate or unintended changes are both subject to continual engagement with real people, either those benefiting from gentrification because they purchased property when its value was low, or those renters who are now being displaced.

A cultural and racial transition occurs disturbing the comfort of what had once been a familiar place. Regardless of its socio-economic profile it was 'knowable' and enjoyed by a relatively homogeneous community. Adding to the curious dynamics of urban regeneration in Fort Greene are public policies like mortgage deductibility, Brooklyn's increasing hip image, and (after the above article was written), major projects like the Atlantic Yards redevelopment, attracting a professional basketball team, the Nets. The broader greening strategy of New York City, and the significant decline in major crime—broad based initiatives beyond the reach and control of Brooklyn— have nevertheless contributed to the Borough's attraction along with its being an easy subway ride into Manhattan.

Ironically, given the arrival of a professional sports team, Brooklynites can be excused for maintaining a sense of caution when it comes to this change in their community. Under Robert Moses in the post-war period, urban renewal became a vehicle for the radical transformation of neighborhoods throughout greater New York. Bridges and roads were built and in some cases entire communities, perceived to be social and economic failures, were relocated or simply disappeared.

Brooklyn's beloved baseball team, the Dodgers, was a victim of larger forces including Moses' manipulation of New York's political and planning process. Brooklyn's decline in the 1950s as its more affluent residents left for the suburbs, was the pretext for a Moses plan to build a new ballpark, but his interference finally provided Walter O'Malley with an excuse to move the team to Los Angeles, where, in a further ironic twist, that city betrayed another community by removing the Hispanic residents of Chavez Ravine to build what is today Dodger Stadium.

Both processes—ecological restoration and urban regeneration—are about change, either evolutionary or managed. The human dimension at the center of the latter however requires an approach purposefully different from that attending ecological restoration.

(c) Case 3: Natural Regeneration: adapted from Scrivener 2012

Janice Gilbert, a wetlands ecologist, is Ontario's specialist when it comes to controlling Canada's worst invasive plant, *Phragmites australis, or* Phrag.

Scrivener describes Phrag, in almost apologetic tones as "…. pretty with its seed heads waving like feathery pennants in the Big Creek wetland, which drains into Lake Erie."

But she acknowledges it's a vicious invader putting animal life and species such as the spotted turtle and Fowler's toad at risk of disappearing. "Plant diversity", she says, "Vanishes. A wetland usually abundant in cattails, bulrushes and sedges becomes a deceptively beautiful monoculture."

Getting rid of it is tough and methods include cutting or burning in specific sites, and even spraying chemical herbicides. While the latter method runs contrary to Gilbert's instinct, she says, "I don't see any other way." Native plants return and flourish when Phragmites are killed off by a herbicide.

Habitats will be lost, Gilbert concludes, unless humans choose to be a corrective force.

(d) Case 4: Built Regeneration: adapted from Doran 2012

For JJ Jegede, a 26-year-old British Olympic long-jump hopeful, the changes in the place in which he grew up were extraordinary. East London was a dodgy part of the city, somewhere you didn't want to go. "Now you see thousands of people in East London coming 'round and seeing it as tourist venue. That's amazing," he said.

But, writes D'Arcy Doran, "Some residents say the regeneration is forcing them to leave. Rob Williams, a 43-year-old train operator, lives in a council-owned tower block overlooking the Olympic Stadium that will be torn down after the Games. 'The worst part by far is that the impact of the Olympics on land prices has meant that the council is relocating us, one by one, destroying what was a vibrant, happy community,' Williams told the Evening Standard."

Places, like Williams' Tower, however date back to London's post-war social housing schemes. They had reached a state of disrepair and were so depressing, said former British housing minister Nick Raynsford, that the best answer appeared to be their replacement with a new mixed development".

(e) Lessons

1. Natural regeneration can and often must be backward looking, though in an era of climate change some strategies must look towards the future in envisioning appropriate planting regimes:

   Once the ponds turned turbid, it wasn't enough to just replant and restock. You had to get them back to their original, clear state.

   …whereas built regeneration is never a return to some original clear state, it does respect history and its sometimes better models.

2. Natural regeneration respects the reality of dynamic systems:

   … the ponds behaved according to a branch of mathematics called 'dynamical systems', which deals with sudden changes. Once you reach a tipping point, it's very difficult to return things to how they used to be.

   …whereas built regeneration also confronts situations in which sudden changes have reached a tipping point from which a return to even a semblance of the past is not possible.

3. Natural regeneration often means the destruction of unwelcome communities:

   …. Mr. Scheffer was able to figure out that to fix the ponds, he had to remove the fish that thrive in the turbid water.

   … whereas built regeneration often occurs in places with social challenges of homelessness, aggressive panhandling, and addiction (nearby methadone clinics for instance) but for which 'destruction' (removal, imprisonment, containment, etc.) contradicts ethics and human rights.

4. Natural regeneration often disturbs apparently tranquil and aesthetically pleasing sites:

   Phrag, as she calls it—is pretty with its seed heads waving like feathery pennants in the Big Creek wetland… A wetland usually abundant in cattails, bulrushes and sedges becomes a deceptively beautiful monoculture.

   …whereas built regeneration has generally left low density car—dependent sites alone. It had enough other challenges. Now however it is at least addressing the future of those monocultural places, though such suburban single-type housing developments are preferred by many of their residents.

5. Natural regeneration often requires radical, some might say violent solutions:

   Mechanical control methods such as cutting or burning can control the reed in specific sites, but spraying chemical herbicides seems to be the most effective tools and sometimes have less impact on the ecosystem.

   ...whereas built regeneration which once adopted radical, even violent means to achieve its ends, now favors a more inclusive approach. The former approaches however often remain popular in totalitarian settings.

6. Natural regeneration requires rational, scientific, engineering intervention by humans:

   There's no evidence to support the argument of letting Mother Nature sort it out. "We need to be a corrective force."

   ...whereas built regeneration has an equivalent to Mother Nature in the unhindered market place—whatever that contested concept might mean! More often it is about humans as a corrective force whether their role is intentional, incremental, or just muddling through.

(f) Conclusions. Regeneration is a Process of Human Intervention

Natural regeneration is about humans intervening in natural environments. Such places, though they obey an evolutionary pattern long pre-dating us, are now increasingly subject to our influence. Humans today have a direct responsibility for the health, resilience, and future of nature, either as something that will continue to provide a variety of ecosystem benefits, or as a totality in decline. Our challenge is to figure out what is an appropriate state—left alone, returned to a past stasis, or, though retaining that past diversity, developing new configurations alert to a changing climate.

Many jurisdictions are looking south for clues as to the planting regimes and forestry in those more temperate climate zones, but which are likely to characterize their more northern locales over the next 25 years.

Built regeneration is about intervening in places where humans live, work, relax, perform, create and all the other marvelous things we do. Returning those in apparent decline to a healthy, more vibrant and fulsome life however is subject to the many points of view held by humans. Like natural systems we also need diversity, not only to maintain human comfort in a changing climate but for reasons ranging from improved public health, to increased opportunities for all.

Many jurisdictions are recognizing the megapolitan character of successful cites and the need to address issues of gentrification, likely dislocation, but also regional mobility for those living in the more expansive geographic territory of the city.

As a reminder, on the one hand:

- "I don't mind when a community changes", said the man, a photographer with a salt-and-pepper beard who gave his name only as Gregory. "But the way it's changing, it's not changing for me and you. It's changing for them."
- Rob Williams, a 43-year-old train operator, lives in a council-owned tower block overlooking the Olympic Stadium that will be torn down after the Games. "The worst part by far is that the impact of the Olympics on land prices has meant that the council is relocating us, one by one, destroying what was a vibrant, happy community."

But on the other hand:

- "You can't say it's not changing for the better for us," Mr. Avery said. "A lot of black folks who owned homes here took the money and ran, moved to Staten Island or down South or wherever. I live here, and it's changing for the better for me."
- JJ Jegede, a 26-year-old Olympic long-jump hopeful, said the transformation of the area where he grew up has been incredible. "As a kid growing up, East London wasn't a good place to go," he said. "Now you see thousands of people in East London coming 'round and seeing it as tourist venue. That's amazing."

Successful regeneration has a pragmatic purposefulness owing little to ideology, self-interest, or old-time rural/urban dichotomies. It is able as necessary to cut at right angles through all paradigms and perspectives as it fashions solutions not just to save what we value but to create more of what we need. It's about listening to, managing, and engaging multiple points of view.

One small example suffices. Zoning was introduced as one of the last measures of the progressive era in the early 20th century to prevent the placement of noxious industries near people's houses. The off-shoring of production in recent years together with what is known as 'clean capitalism' have made this a receding concern and the city of Houston for instance has relied on a 'libertarian' absence of zoning policy. One might see this as a clear left-right issue (zoning on the left, no zoning on the right). However Mike Davis has painted a progressive picture of Latinos in Los Angeles bypassing restrictive zoning to revitalize once downtrodden, bleak neighborhoods by a combination of front lawn bistros, repair shops, millinery operations, and other one-off businesses all violating local codes and yet all contributing to a regenerating place (Davis 2001).

A regeneration corps of citizens, volunteers and workers of all ages, perhaps modeled on FDR's groundbreaking Civilian Conservation Corps, are the potential tactical, and pragmatic, urbanists and countryside dwellers of the future, tackling the nuances and subtleties of our common challenge. As well those practicing in built and natural environments including those with professional legal status for design and accreditation, those responsible for maintaining and tweaking those systems over their lifetime, and those who work, live and provide stewardship within these places, must collaborate, converse, and construct with attentiveness to a regeneration imperative.

References

Birkeland, J. 2009. Positive Development: From Vicious Cycles to Virtuous Cycles through Built Environment Design. Earthscan, London.

Booz Allen Hamilton, Strategy & Business, no. 46, 2007 (from Booz Allen Hamilton, Global Infrastructure Partners, World Energy Outlook, OECD, Boeing, Drewry Shipping Consultants, U.S. Department of Transportation).

Cirillo, C. and L. Podolsky. 2012. Health, Prosperity and Sustainability: The Case for Green Infrastructure in Ontario. Green Infrastructure Ontario Coalition and Ecojustice, Toronto.

Cunningham, S. 2002. The Restoration Economy: The Greatest New Growth Frontier. Berret-Koehler, San Franciso.

Davis, M. 2001. Magical Urbanism. Verso.

Doran, D. 10 August 2012. A Never-ending Story London 2012: City looks to its Olympic legacy as Games wind down. Toronto Star.

Englander, J. 2012. High Tide on Main Street. The Science Bookshelf, Boca Raton.

Guterl, F. 20 July 2012. Search for Clues to Calamity. New York Times Green Infrastructure Ontario

Harvey, F. 9 October 2008. Staying on Course in a Tougher Climate. Sustainable Business, Financial Times Special Report.

Jacobs, J. 2004. Dark Age Ahead. Random House, New York.

Kotkin, J. 2012. The Rise of the Great Plains: Regional Opportunity in the 21st Century, 2012 Project Sponsor, Office of the President, Texas tech. University; 96.

Kuhlberg, M. 1974. Damaged and Efficient Landscapes in Rural Southern Ontario. Ontario History 67(1).

LaRusic, E. May 2014. More than Just Tax Breaks. Novae res Urbis Vol. 17 No. 20 14.

Mallach, A. and L. Brachman. 2013. Regenerating America's Legacy Cities, Policy Focus Report, Lincoln Institute of Land Policy 13-14.

Marx, L. 1964. Machine in the Garden: Technology and the Pastoral Ideal in America. Oxford University Press, London.

Newman, A. 31 August 2001. Fort Greene, Prosperity Is Bittersweet; Some Blacks Reap Profits As Others Lament Change. New York Times.

O'Connor, M. 29/30 March 2014. Green Dean. Financial Times; 23.

Partridge, M. and J. Clark. 2008. Our Joint Future: Rural-Urban Interdependence in Twenty-First Century Ohio. White Paper for Greater Ohio Policy Center and Brookings Institution. Columbus, Ohio.

Rosenthal, J. 2014. Superstorm Sandy and the Age of Preparedness. Harvard Design Magazine 37, Harvard University Graduate School of Design; 33.

Saunders, D. 22 December 2007. The Hush-Hush Regreening of Europe, Globe and Mail.

Scriver, L. 11 August 2012. *Phragmites austrailis* is Canada's worst invasive plant, Toronto Star

Shorto, R. 13 April 2014. Water Works, New York Times Magazine.

Sustainable Prosperity. March 2014. The Importance of Natural Capital to Canada's Economy. Policy Brief.

Example of a Green Building

McMaster Innovation Park (MIP): CANMET – MTL Building

Focusing on innovative research and entrepreneurship, MIP provides a collaborative and collegial environment for academic, government and industry research and development. The innovation park was a former industrial manufacturing site with several buildings that have been re-purposed for new uses and provides a classic example of community regeneration. The Park is in close proximity to major arterial roads, rail lines, businesses and McMaster University which provides support for research initiatives. The CANMET-MTL building is a dedicated research facility housing the Canadian Government's Materials Technology Laboratory. The operation focuses on metals and materials fabrication, processing and evaluation.

Location: Hamilton, Ontario
Size: 11,671 m^2 or 165,000 ft^2
Certification: LEED Canada Platinum

Green Technologies and Features

Site

- Brownfield re-development and remediation

Materials

- Local sourcing of most building materials
- Recycled and rapidly renewable materials for interior finishes
- High albedo (reflective) roof and envelope

Water Conservation

- Rain water harvesting
- Grey water applications
- Storm Water Management

Energy

- Sixty-five percent improvement over the MNBC
- Solar Thermal applications
- Connection to District Energy System including ground-source geo-energy system

Ventilation

- Solar Wall ventilation and preheat applications

Picture and Description, Courtesy: Tony Cupido

Chapter 4

Collaborative Problem Solving

Collaborative problem solving is a method where individuals work together to find a mutually acceptable solution to a given matter. The process clarifies and redefines a perceived problem, identifies novel alternatives and seeks our overlapping interests. No one party dominates the process, and all parties benefit. This is commonly recognized as a win/win approach to resolving conflicting needs.

Fisher and Ury (1983) explain that a good agreement is one which is wise and efficient, and simultaneously improves the relationships among the parties. Such agreements meet the parties' interests and as a result are practical and enduring.

The process differs from negotiations that take the form of positional bargaining. In positional bargaining each party articulates their position on an issue. The parties then bargain from their separate opening positions to eventually agree on one of the position. Fisher and Ury, however, argue that positional bargaining generally does not produce good agreements. The agreements reached in such a process tend to neglect the parties' interests. It promotes inflexibility and so harms the relationships among the parties'.

Principled negotiation provides a better way of reaching good agreements. The Fisher and Ury process of principled negotiation can be used effectively on almost any type of dispute. Their four principles are: (1) separate the people from the problem; (2) focus on interests rather than positions; (3) generate a variety of options before settling on an agreement; and (4) insist that the agreement be based on objective criteria. Such a negotiation is based on open communication designed to reach agreement by illuminating the interest that all stakeholders share. Also referred to as collaborative problem solving, the procedure requires the use of listening skills, assertion skills and uses the conflict resolution method.

Steps Associated with Collaborative Problem Solving Model
(modified from Willihnganz 2001)

1. Separate people from issues.
2. Define the problem in terms of needs not solutions.
3. Brainstorm possible solutions.
4. Select the solution that will best meet all parties needs and check possible consequences.
5. Use objective criteria.
6. Plan who will do what, where, and by when.
7. Implement the plan.
8. Evaluate the problem-solving process and at a later date how well the solution worked out.

Step 1: Separate People from Issues

People are often personally involved with their position on a particular issue and tend to interpret responses to those positions personally. Separating the individuals from the issues allows the parties to discuss the issues in a non-threatening manner thereby protecting their relationships. It facilitates the identification of the substantive matters, by separating the person from the substance.

Personal conflicts take various forms. Since most conflicts arise from differences in the interpretations of facts, central to resolution are exercises that enable each side to understand the other's viewpoint. One method is for each party to put themselves in the other's place. The parties should be aware that some individuals assume that their worst fears will be a result of the actions of the other party. Each participant should attempt to find proposals that would appeal to the other side. The more that the parties are involved in the process of finding mutually beneficial solutions, the more likely they are to support the outcome.

Negotiation can be a frustrating process. People react negatively when they feel that their interests are threatened. Basic skills in defusing intense emotions involve acknowledging them and understanding their source. Dismissing fear or anger often provokes stronger emotional responses. Symbolic gestures such as apologies or an expression of understanding can help to defuse strong emotions.

Communication is a central source of people problems. The parties often do not listen to each other, but instead concentrate on how they will respond and deflect the other's position. To address the communication problem, the parties need to overtly understand the importance of active listening. The process involves listeners giving the speaker their full attention, occasionally summarizing the speaker's points to confirm their understanding. This does not mean that understanding the other's case is the same as agreeing with it. It means considering each other as partners in negotiation rather than as adversaries. The primary characteristics of partnership success as identified by Mohr and Spekman (1994) are commitment, coordination, and trust; communication quality and participation; and the conflict resolution technique of joint problem solving.

Step 2: Define the Problem in Terms of Needs not Solutions

For a win/win outcome the problem is defined in terms of needs and not proposed solution. To uncover the diversity of needs a useful exercise is to delve into why a party prefers a different outcome than the one proposed. Good agreements focus on the parties' interests, rather than their positions. As Fisher and Ury explain, "Your position is something you have decided upon. Your interests are what caused you to so decide" [p. 42]. Defining a problem in terms of positions means that at least one party will 'lose' the dispute. When a problem is defined in terms of the parties' underlying interests it is often possible to find a solution which satisfies the interests of the involved parties.

The first step is to identify the parties' interests regarding the issue at hand. This can be done by asking why they hold the positions they do, and by considering why they don't hold some other possible position. Of fundamental importance, however, is that all participants will share some basic interests or needs, such as the need for security, economic well-being, and other quality of life matters.

An interest is a person's concern, desire or goal. While it is often confused with a position or solution, it is rather, the underlying need that must be met if agreement is to be reached. It usually requires clarifying because it is not always immediately evident, and the interest provides the motivation to seek resolution through solutions.

Once the parties have identified their interests, they must discuss them together. If a party wants the other side to take their interests into account, that party must explain their interests clearly. The other side will be more motivated to take those interests into account if the first party shows that they are paying attention to the other side's interests. Discussions should look forward to the desired solution, rather than focusing on past events. Parties should keep a clear focus on their interests, but remain open to different proposals and positions.

Seeking to understand how the other side sees the situation may not only help us see potential solutions that will meet many of our needs, but may also allow us to revise how we see the problem. The result is that the area of conflict may actually be found much smaller than originally perceived, bearing in mind that perception is a personal interpretation based on one's experiences, modified by one's values, fears, desires, and risk tolerance, and importantly, unique to everyone. It is differences in people's perceptions that cause conflict, so that understanding how people perceive themselves and the world around them is the key to understanding their behavior and will help open ways to finding solutions. Through this process of sharing perspectives, we are able to begin identifying the issues that will need to be addressed.

Step 3: Brainstorm Possible Solutions

Brainstorming is the rapid generation and listing of solution ideas without clarification and without evaluation of their merits. The rules are: (1) generate a lot of ideas; (2) avoid criticizing any of the ideas; (3) attempt to combine and improve on previously articulated ideas; and (4) encourage the generation of 'wild' ideas. These rules can be viewed as assigned goals because individuals are asked to strive to follow them (Litchfield 2008). In this exercise, creative thinking is valued and the experience demonstrates that working together can be more creative in dealing with a common concern than working alone.

Fisher and Ury identify four obstacles to generating creative options for solving a problem. Parties may decide prematurely on an option and so fail to consider alternatives. The parties may be intent on narrowing their options to find the single answer. The parties may define the problem in win-lose terms, assuming that the only options are for one side to win and the other to lose. Or a party may decide that it is up to the other side to come up with a solution to the problem.

There are techniques for overcoming these obstacles and generating creative options. First it is important to separate the invention process from the evaluation stage. The parties should come together in an informal atmosphere and brainstorm for all possible solutions to the problem. Wild and creative proposals are encouraged. Brainstorming sessions can be made more creative and productive by encouraging the parties to shift between four types of thinking: stating the problem, analyzing the problem, considering general approaches, and considering specific actions. Parties may suggest partial solutions to the problem. Only after a variety of proposals have been made should the group turn to evaluating the ideas. Evaluation should start with the most promising proposals. The parties may also refine and improve proposals at this point.

Participants can avoid falling into a win-lose mentality by focusing on shared interests. When the parties' interests differ, they should seek options in which those differences can be made compatible or even complementary. The key to reconciling different interests is to "look for items that are of low cost to you and high benefit to them, and vice versa" [Fisher and Ury 1982, p. 79]. Each side should try to make proposals that are appealing to the other side, and that the other side would find easy to agree to. To do this it is important to identify the decision makers and target proposals directly toward them. Proposals are easier to agree to when they seem legitimate, or when they are supported by precedent. Threats are usually less effective at motivating agreement than are beneficial offers.

Rules for Brainstorming[1]

Postpone and Withhold your Judgment of Ideas

Do not pass judgment on ideas until the completion of the brainstorming session. Do not suggest that an idea will not work or that it has bad side-effects. All ideas are potentially good so do not judge them until afterwards. Avoid discussing ideas, which includes not criticizing and not complimenting ideas.

Ideas should be put forward both as solutions and also as a basis to spark off solutions. Even seemingly foolish ideas can spark off better ones. Therefore do not judge the ideas until after the brainstorming process.

Note down all ideas. There is no such thing as a bad idea.

Evaluation of ideas takes up valuable brain power which should be devoted to the creation of ideas.

Maximize your brainstorming session by only spending time generating new ideas.

Encourage Wild and Exaggerated Ideas

It is much easier to tame a wild idea than it is to think of an immediately valid one in the first place. The 'wilder' the idea, the better. Shout out bizarre and unworkable ideas to see what they spark off. No idea is too ridiculous. State any outlandish ideas. Exaggerate ideas to the extreme.

Use creative thinking techniques and tools to start your thinking from a fresh direction. Use specialist software such as Innovation Toolbox to stimulate new ideas more easily.

Quantity Counts at this Stage, not Quality

The more creative ideas a person or a group has to choose from, the better. If the number of ideas at the end of the session is very large, there is a greater chance of finding a really good idea. Keep each idea short, do not describe it in detail—just capture the essence of the idea. Brief clarifications can be requested. Think fast, reflect later.

Go for quantity of ideas at this point; narrow down the list later. All activities should be geared towards extracting as many ideas as possible in a given period.

Build on the Ideas Put Forward by Others

Build and expand on the ideas of others. Try and add extra thoughts to each idea. Use other people's ideas as inspiration for your own. Creative people

---

[1] Visit www.brainstorming.co.uk, internet and computer resources for creativity and brainstorming. ©1999 Infinite Innovations Ltd. All rights reserved.

are also good listeners. Combine several of the suggested ideas to explore new possibilities.

It is just as valuable to be able to adapt and improve other people's ideas as it is to generate the initial idea that sets off new trains of thought.

Every Person and Every Idea has Equal Worth

Every person has a valid view point and a unique perspective on the situation and solution. We want to know yours. In a brainstorming session you can always put forward ideas purely to spark off other people and not just as a final solution. Please participate, even if you need to write your ideas on a piece of paper and hand it out. Encourage participation from everyone.

Each idea presented belongs to the group, not to the person stating it. It is the group's responsibility and an indication of its ability to brainstorm if all participants feel able to contribute freely and confidently.

Step 4: Select the Solution (or Combination of Solutions) that will Best Meet all Parties' Needs

This process begins by one party asking what solutions the others would favor in the resolution of the problem, and also making it clear which alternatives are best for that party. Some of the choices may well coincide. Jointly decide on one or more of the alternatives. It is usually the case that if the needs were well defined at the start, several of the same alternatives will be selected by the parties. Figure 4.1 presents elements necessary for collaboration to advance.

Ensure that all parties are satisfied with the solution so there will be motivation to move towards implementation. Consensus, rather than voting, is the most appropriate decision-making process to use in this situation. Consensus is possible when a process of free and open exchange of ideas

Figure 4.1. Illustration of collaborative elements using a collaborative word cloud.

permits agreement to be reached. Each individual's concerns are heard and understood and a sincere attempt to take them into consideration is made. The conclusion may not reflect the exact wishes of any member, that is consensus may not be concurrence in that it does not eliminate concerns of the parties, but it is an approach that can be agreed upon.

With a comprehensive list of brainstormed ideas and a mutually agreeable objective criteria, the next exercise is to evaluate the options and move toward agreement on an approach that will meet as many of collective needs and interests as possible.

Techniques for Narrowing the Field of Options (after Windel and Warren)

Thumbs Up/Thumbs Down

This method provides a general sense of the parties' views on any particular item. Using the objective criteria, the parties go through the list of items and give a thumbs up, thumbs down, or thumbs neutral sign. Some items will obviously not meet the objective standard and can be eliminated with unanimous thumbs down.

Using Stars

Star the items that the group thinks are best.
Can any of the favored options be reworked to create even better options?

Combining Items for Mutual Gain

Some items may partially meet the objective criteria. Can some of these ideas be combined to create agreements for mutual gain? Can they be combined to actually meet more of the parties' needs and interests, thereby expanding the total pie?

Throughout the process, participants need to continually refer back to the identified interests and the objective criteria to make sure they are meeting as many of their collective interests as possible. Always be asking the question, "Is this the best we can do?"

Step 5: Use Objective Criteria

When interests are directly opposed, the parties should use objective criteria to resolve their differences. Decisions based on reasonable standards make it easier for the parties to agree and preserve their good relationship.

The first step is to develop objective criteria. Usually there are a number of different criteria which could be used. The parties must agree which criteria is best for their situation. Criteria should be both legitimate and practical. Scientific findings, professional standards, or legal precedent are possible sources of objective criteria. Some popular criteria are product or service quality, followed by delivery, price/cost, manufacturing capability, management, technology, research and development, finance, flexibility, reputation, relationship, risk, safety, and environment (Ho et al. 2010).

Each issue should be approached as a shared search for objective criteria. Clarify the reasoning behind any party's suggestions. Further each party must keep an open mind and be reasonable, and willing to reconsider their positions when there is reason to.

Windle and Warren (nd) discuss the benefits of using Objective Criteria

1. Protects the relationship from a contest of wills;
2. Allows the parties to use the time more effectively, focusing rather on standards and solutions rather than on defending their positions;
3. Enables parties to alter their perceptions without 'losing face';
4. Enables parties to strive toward mutual fairness and decisions that are in the best interest of the child;
5. Creates agreements that are fair and wise.

Step 6: Plan who will do what where, and by when

This element of collaborative decision making involves detailing the pragmatic aspects of the solution and thinking collectively through any unforeseen consequences. Once the details have been worked out it is usually helpful to document them. At this point in the process it should be clear that all parties are willing to make joint decisions and coordinate the appropriate plans to meet needs.

Step 7: Implement the Plan

Complete the action steps on schedule as a measure of good faith. If there is failure to do so, an assertion message followed by reflective listening may be appropriate.

The message conveyed is that the parties have the ability and the power to change their behaviors in ways that will enhance their lives and their relationship. Commitment is expressed in actions in addition to words.

Step 8: Evaluate the Problem-Solving Process and at a later date how well the Solution Worked Out

Evaluate the process—what each party most liked and most disliked about the process, and what each can do better the next time. Set a time for evaluation of the solution (after it has had a chance to be implemented). If it is not working, it needs to be corrected, or a new one needs to be established. If it is working well it needs to be celebrated in some way.

At this point in collaboration, all parties demonstrate honesty and integrity by discussing their interactions. They are not locked into any solution. If it doesn't turn out as good as anticipated they have the power to make it better.

Sample Mediation Ground Rules

1. We will take turns speaking and not interrupt each other.
2. We will call each other by our first names, not "he" or "she".
3. We will not blame, attack, or engage in put-downs and will ask questions of each other for the purposes of gaining clarity and understanding only.
4. We will stay away from establishing hard positions and express ourselves in terms of our personal needs and interests and the outcomes we wish to realize.
5. We will listen respectfully and sincerely try to understand the other person's needs and interests.
6. We recognize that even if we do not agree with it, each of us is entitled to our own perspective.
7. We will not dwell on things that did not work in the past, but instead will focus on the future we would like to create.
8. We will make a conscious, sincere effort to refrain from unproductive arguing, venting, or narration, and agree at all times to use our time in mediation to work toward what we perceive to be our fairest and most constructive agreement possible.
9. We will speak up if something is not working for us in mediation.
10. We will request a break when we need to.
11. We will point out if we feel the mediator is not being impartial.

The National Center on Dispute Resolution in Special Education, http://www.directionservice.org/cadre/grs.cfm

---

When the Other Party Won't Use Principled Negotiation

Sometimes the other side refuses to budge from their positions, makes personal attacks, seeks only to maximize their own gains, and generally refuses to partake in principled negotiations. Fisher and Ury describe three approaches for dealing with opponents who are stuck in positional bargaining. First, one side may simply continue to use the principled approach. The authors point out that this approach is often contagious.

Second, when the other side attacks, the principles party should not counter attack, but should deflect the attack back onto the problem. Positional bargainers usually attack either by asserting their position, or by attacking the other side's ideas or people. When they assert their position, respond by asking for the reasons behind that position. When they attack the other side's ideas, the principle party should take it as constructive criticism and invite further feedback and advice. The principled party should use questions and strategic silences to draw the other party out.

When the other party remains stuck in positional bargaining, the one-text approach may be used. In this approach a third party is brought in. The third party should interview each side separately to determine what their underlying interests are. The third party then assembles a list of their interests and asks each side for their comments and criticisms of the list. The third party then takes comments and draws up a proposal. The proposal is given to the parties for comments, redrafted, and returned again for more comments. This process continues until the third party feels that no further improvements can be made. At that point, the parties must decide whether to accept the refined proposal or to abandon negotiations.

References

Fisher, R. and W. Ury. 1983. Getting to Yes: Negotiating Agreement Without Giving In. Penguin Books, New York.

Ho, W., X. Xu and P.K. Day. 2010. Multi-criteria decision making approaches for supplier evaluation and selection: A literature review. European Journal of Operational Research 202: 16–24.

Litchfield, R.C. 2008. Brainstorming reconsidered: A goal-based view. Academy of Management Review 33: 649–668.

Mohr, J. and R. Spekman. 1994. Characteristics of partnership success: Partnership attributes, communication behavior, and conflict resolution techniques. Strategic Management Journal 15: 135–152.

Windle, R. and S. Warren. Collaborative Problem Solving: Steps in the Process http://www.directionservice.org/cadre/section5.cfm/contents.cfm (accessed 7/2012).

Willihnganz, N. 2001. http://willihnganz.disted.camosun.bc.ca/collaborativeps.htm.

Example of a Green Building

Evergreen Brickworks (Torza 2011)

This complex is located in Toronto's Don Valley and was the former home of the Don Valley brickworks which manufactured bricks that built many landmark buildings across Canada for over 100 years. When it closed down in the late 1980s, the 42 acre site had a damaged ecosystem that was contaminated with by-products of the brick-making process. The site contained 16 industrial heritage buildings including masonry, steel and wood-framed structures built between the 1880s and 1960s. The complex is located in the floodplain of the Don River where regional storm event flood waters would exceed several metres.

Evergreen, a Canadian non-profit organization with stewardship of the complex, has transformed it into Canada's first large-scale community environmental centre. Several of the existing structures have been renovated and a new innovative building, known as the Centre of Green Cities, has been constructed. It is a classic example of adaptive reuse with most of the buildings repurposed to house new programming planned for the site. Evergreen hosts many conferences and courses which gives participants the opportunity to go through most of the dynamic and interactive spaces that showcase how nature is central to all aspects of urban life.

Location: Toronto, Ontario
Size: 16,537 m² or 198,708 ft²
Certification: LEED Canada Platinum NC

Green Technologies and Features

Materials

- Building insulation includes a soy-based spray foam as well as locally-manufactured mineral wood batts.
- Local materials with high recycled content and sustainably harvested wood products.
- Recyclable carpet tile in selected locations.

Energy

- Electrical power and natural gas supplied by Bullfrog power
- Energy savings relative to the Model National Energy Code for Buildings = 50%
- Centralized and distributed sensors and controls for monitoring and managing energy use both on-site and remotely.

Ventilation

- Heating, cooling and ventilation consist of active and passive systems, including natural ventilation though operable windows and thermal chimneys. Active systems, fed by high efficiency boilers, include radiant perimeter and in-floor heating and kitchen and ventilation exhaust heat recovery. Three solar thermal chimneys are fitted with louvres with motorized dampers and have acoustically-insulated sheet metal elbows to reduce noise transfer between floors.
- The new Centre for Green Cities building is constructed of concrete with distribution ducting located along the centre structure and routed to the exterior through the hollow core slab, pre-cooling the thermal mass.
- E-mails are sent to occupants advising them to take action to open windows if optimal outdoor temperatures are favourable.
- "Free" night cooling is utilized when available.

Other Innovations and Technology

- Utilization of a web-enabled dashboard that provides easy access to all pertinent data for public engagement and education.
- On-line and hand-held tools for education, interpretation and civic engagement.
- Photo-monitoring stations across the campus allow visitors to help Evergreen communicate through images.
- Targeted Social media and use of Blogging, Twitter and Facebook have attracted new followers.

Chapter **5**

Remedial Action Plans for Great Lakes Regeneration

Introduction

In 1909, there was an event that demonstrated respectful wisdom, that water does not abide by political boundaries. The signing of the Boundary Waters Treaty by Britain, on behalf of Canada, and United States was a landmark event to protect the shared waters of the two nations. It has provided the framework for cooperation on questions relating to air and water pollution and the regulation of water levels and flows. It created the International Joint Commission (IJC) that undertakes investigations of specific issues, or monitors situations, when requested to do so by Governments. IJC recommendations concerning pollution in the Great Lakes served as the basis for the Governments to create the Great Lakes Water Quality Agreement, which was signed by the Prime Minister of Canada and the President of the United States on April 15, 1972 (United States and Canada 1972).

In November 22, 1978, a revision to the 1972 Agreement was signed. It provided new programs and more stringent goals to eliminate pollution, particularly, persistent toxic substances from the lakes. The concept of a Great Lakes Basin Ecosystem was adopted, which recognizes that water quality depends on the interplay of air, land, water and living organisms, including humans, within the system. This action led to more comprehensive assessments of the Great Lakes cleanup effort.

In November 1987, the Governments signed a Protocol which aimed to strengthen the programs, practices and technology prescribed in the 1978 Agreement and to increase accountability for their implementation. The governments made the commitment to restore and maintain the chemical, physical and biological integrity of the waters of the Great Lakes Basin Ecosystem (United States and Canada 1972). Remedial Action Plans (RAPs) were described in the 1987 amendment of the Agreement under Annex 2.

The United States and Canada revision in 1987 of the Great Lakes Water Quality Agreement specifically calls for the development of RAPs at Areas of Concern (42 in total at that time) where ecosystem deterioration is particularly pronounced (United States and Canada 1987). A RAP is a tool through which governments and concerned citizens can restore and protect 'beneficial uses' (14 of which are specified in the agreement). The two federal governments directed their national environmental agencies to work in cooperation with state and provincial governments and with local communities to jointly develop and cooperatively implement the RAPs.

The restoration experiments, as suggested by Sproule-Jones (2002), promise a way in which resource users, regulators, and those in an interest in restoring the local ecosystem can collaborate towards a common purpose. They promise to empower local stakeholders to determine their own solutions to ecological degradation, and open new venues for collaboration.

With the assistance of governments, residents in most AOCs formed an advisory council/committee to work with federal/state/provincial technical and scientific experts. These committees typically have or had representatives from diverse community sectors, including, agriculture, business and industry, citizens-at-large, community groups, conservation and environment, education, fisheries, health, labour, municipal governments, native peoples, shipping, tourism and recreation. Engaging stakeholder groups in the plan design minimizes the risk of future polarization (Samy et al. 2003). Advisory Committee participants possess unique knowledge and represent the interests of their particular stakeholder groups. A key premise is that community residents pose important knowledge, and can provide an informed perspective of the social impacts of the decisions (Harris et al. 2003). The importance of involving communities in the management of water resources was one of the strongest and most consistent messages coming forward from a recent conference (Managing Shared Waters 2002). It is a matter of recognizing the value of traditional knowledge and the public's anecdotal and experiential expertise. Good public processes use plan language to communicate clearly, are supported by commitments in institutional workplans, demonstrate clearly how public input will be used, include mechanisms to resolve disputes, provide the community with access to technical experts, celebrate successes and train community leaders.

Stakeholders have been instrumental in helping governments be more responsive to and responsible for restoring uses in AOCs. Further, stakeholders have been the primary catalyst for implementing actions which have resulted in ecosystem improvements. Such broad-based partnerships among diverse stakeholders can best be described as a step towards grassroots ecological democracy in the Great Lakes Basin (Hartig and Zarull 1992). The collective objective is to work with governments and develop a plan to revitalize ecosystem health and implement the plan to achieve agreed-upon targets that indicate when beneficial uses are restored.

A key concept in the RAP process is accountability for action. This is established through open sharing of information, clear definition of problems (including identification of indicators to be used in measuring when the desired state is reached), identification of causes, agreement on actions needed, and identification of who is responsible for taking action. From this foundation, the responsible institutions and individuals can be held accountable for progress (Hartig and Zarull 1992).

Since 1987, incremental progress has been made to restore beneficial uses in the Areas of Concern. Approximately 15 years since the inception of the RAP program, hundreds of kilometres of riparian vegetation and thousands of hectares of wetlands have been rehabilitated (Canada-Ontario 1999). Sediment quality is improving in some locations because of pollution control and some sediment clean-up (IJC 1997). More fish are edible in more places and swimming is again possible in parts of our urban centres for the first time in decades (Krantzberg et al. 1999). Tens of thousands of volunteers are giving their energy to revitalize their homes. Scores of funding partners have collaborated (Environment Canada 1997). Research is being advanced basin-wide on the insidious nature of toxic chemicals. Technologies are emerging to better manage stormwater and wastewater, and contaminated sediment. There have been declines in chemical concentrations in Great Lakes fish (Ontario Ministry of Environment 2001).

Notwithstanding these strides forward, human health is still being compromised by toxic chemicals, particularly for those consuming fish that are contaminated at unsafe levels, and particularly for children exposed to contaminant in utero (Schwartz et al. 1983, Davidson et al. 1995, Jacobson and Jacobson 1996, Lonky et al. 1996). More aggressive action to revitalize the Lakes is essential to protect the health of all their residents (IJC 2003).

More than 33 million people inhabit the Great Lakes Basin, including about a third of Canada's population and 15% of United States' population. The Great Lakes and other lakes and rivers in the Basin provide drinking water to millions. On both sides of the border, the Basin supports multibillion dollar manufacturing, service, tourism and outdoor recreation industries as well as strong maritime transportation systems and diversified agricultural sectors. It provides the foundation for trade between Canada and the United States, equaling 50% of Canada's annual trade with the United States. Degradation of environmental quality directly impairs the viability and vitality of the region. The reliance of the economy on a healthy Great Lakes Basin Ecosystem is unequivocal and the imperative to restore ecosystem health is clear. To achieve sustainability, ecology and economics must be integrated. Sustainability can be defined as a balanced relationship between the dynamic human economic systems and the dynamic, but generally slower-changing ecological systems in which: (1) human life can continue indefinitely; (2) people can flourish; (3) cultures can develop, but within bounds such that human activities do not destroy the diversity, complexity, and function of the ecological life-support system (Costanza 1992).

Evidence of Progress

Stakeholders in various AOCs in the United States and Canada have made considerable investments of time and money, and several well-documented successes are highly visible (IJC 2003). Gurtner-Zimmermann (1995) notes that the commitment of individuals who participate in the RAP process, local support for the RAP goals, and the scientific basis and sound analysis of environmental issues contribute to the positive outcomes.

Major successes include Collingwood Harbour and Severn Sound in Ontario, where conditions have improved to the point that these locations are no longer considered to be Areas of Concern. Spanish Harbour in Ontario and Presque Isle Bay in Pennsylvania are now recognized as in a stage of recovery due to completion of all selected remedial actions, while monitoring continues to measure recovery of beneficial uses.

Other examples of successes include the removal of over 1.3 million cubic yards (1 million cubic metres) of sediment contaminated with polychlorinated biphenyls (PCBs) from the Kalamazoo River, Manistique River, Maumee River, Rouge River, Saginaw River, Saginaw Bay, and St. Lawrence River. Approximately Can\$270 million and at least US\$3 billion has been invested

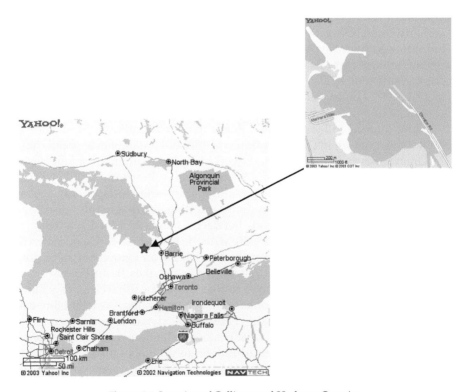

Figure 5.1. Location of Collingwood Harbour, Ontario.

over the last 10 years to improve the condition of wastewater infrastructure in various AOCs (IJC 2003).

But even with these successes, the chemical, physical, and biological integrity of the Great Lakes Basin ecosystem remains threatened. Lack of resources and lack of inter-program coordination and cooperation impede progress (Gurtner-Zimmermann 1995). In some AOCs, environmental problems remain ill-defined both in terms of the magnitude of degradation and the societal costs to either maintain the status quo or undertake adequate remedial actions (IJC 2003).

Differences in the local context of the plans have resulted in a diversity of individual planning and implementation experiences. Notwithstanding this diversity, the motivation and political clout of RAP participants are strongly intervening factors. Resource input from upper levels of government, in particular financial commitment for plan implementation, is also a necessary ingredient for progress due to the RAPs' weak regulatory and institutional framework (Gurtner-Zimmermann 1996).

This chapter now explores the elements that fostered successful cooperative and collaborative initiatives and sustain the objectives of a community engaged in cleaning up its harbour. The discussion is meant to illuminate strategies for revitalizing place-based efforts to restore ecosystem quality in the Areas of Concern.

The RAP process clearly embraces the ecosystem approach. Here, the ecosystem approach is based on the man-in-system concept rather than a system-external-to—man concept (IJC 1978), where the ecosystem is composed of the interacting elements of water, air, land and living organisms including man. While Lee et al. (1982) discuss several variants of the ecosystem approach, most share a focus on the responsiveness of ecological systems to natural and human activities, and a readiness to strike a programmatic compromise between detailed understanding and more comprehensive holistic meaning. This flexible pragmatism is perhaps the most productive feature for addressing Great Lakes environmental problems, and was reflected in the manner in which the Collingwood Harbour RAP was developed and implemented.

What follows is a description of a lesson learned in successful RAP experiments.

Lesson One—Leadership

Engage local leaders who are committed to their community and can affect change. One view, one voice. When leadership around the RAP table is incomplete, it is time to recruit new members.

In establishing the means for community collaboration, senior leaders with local influence, possessing unique points of interest, were contacted and interviewed. The selection of candidates for the Public Advisory Committee (PAC) was based predominantly (but not exclusively) on identifying decision makers who could affect change within the sector or stakeholder group they

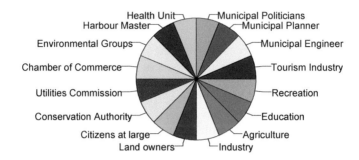

Figure 5.2. Composition of the Collingwood Harbour Public Advisory Committee. The equity of the pies is intended to represent 'one view, one voice'.

represented. This is in keeping with the observation that plan effectiveness will be, in part, a function of the inclusiveness of stakeholder and user representation and goal setting. Inclusivity lends legitimacy, stimulates funding, and can galvanize potentially marginalized but important stakeholders through peer pressure. It has been observed that the wider the scope of stakeholder representation, the stronger the performance of the RAP (Sproule-Jones 2002).

Lesson Two—Consensus on Goals

Articulate clear and meaningful goals early in the process to unite the team. This gives the group the means to overcome conflicts and obstacles during the development and implementation of the Plan.

The leaders constituted the PAC which, in consultation with the community at large, reached consensus on the vision for the future of their harbour or waterfront or river. To gain support for a restoration and rehabilitation strategy, the common vision for the future of the harbour and its watershed was of paramount importance. When there were conflicting opinions on aspects of the restoration plan that threatened further development and implementation of the plan, returning to the shared vision as the fundamental purpose of the RAP enabled the group to re-establish consensus-based decision making.

Lesson Three—Quantifiable Endpoints

Specifying, to the extent possible, quantifiable endpoints or delisting targets that signify success and the achievement of the goals allows the group to recognize progress, prioritize actions and reach consensus on delisting.

To evaluate when or whether the harbour could support the goals and uses formulated by the community, the RAP Team and PAC jointly delineated rehabilitation targets (or delisting criteria) that were, to the extent possible, quantitative and science based. These targets enabled the team to prioritize clean-up efforts, and to measure progress towards the restoration of beneficial uses. This was remarkably helpful for setting priorities for RAP implementation and designing environmental monitoring programs.

The identification of such targets is not deliberately called for in Annex 2 of the GLWQA. Fortunately it is implied (United States and Canada 1987). Under sub-paragraphs 4 (vii) and (viii) Annex 2 calls for:

(vii) a process for evaluating remedial measure implementation *and effectiveness*; and

(viii) a description of surveillance and monitoring process to *track the effectiveness* of remedial measures and the eventual *confirmation of the restoration of uses*. (author=s emphasis).

The ultimate goal of RAPs as described in Annex 2 of the Agreement is to confirm that the beneficial uses have been restored. Since implementation efforts can be prolonged, environmental conditions will presumably change. Further, since our knowledge regarding the threats posed by degraded environmental conditions is evolving, descriptions of the status of beneficial uses and their restoration targets need to be reviewed and updated systematically. For example, human health effects that can be caused by exposure to PCBs or methyl mercury are reflected by fish consumption advisories. These advisories are more restrictive presently than when AOCs were identified. Accordingly, delisting targets must be modified using current data.

Measuring progress toward the delisting targets improves the likelihood that investments in actions result in optimal environmental returns. Many AOC practitioners across the basin presently cannot estimate the degree to which beneficial uses are improving, in part, because the RAPs do not contain targets that help identify when beneficial uses would be considered restored (IJC 2003). The absence of suitable restoration targets and lack of understanding of the current status of beneficial uses represent a real challenge to progress. In times of scarce resources, it is imperative that RAP practitioners affirm that investments in particular active interventions are appropriate and productive (IJC 1988).

Defining delisting targets has proven problematic for many. For example, both the GLWQA and the Ontario provincial water quality guidelines discuss water quality problems in terms of persistence, a term which is not delimited. If a particular water quality objective was not met for a specific location or during a given length of time, the RAP Team and PAC examined the implications for achieving the community-based endpoints (such as the absence of nuisance levels of algae, sufficient oxygen to support aquatic life, aesthetics and enjoyable recreational experiences, fish community health). Persistence, then, was linked to the ability of the water way to support the goals and uses embodied in the vision statement of the PAC.

Tracking the resultant incremental improvement in the restoration of the ecosystem helps identify shortfalls, guide future actions and work plan development, and allows for the prioritization of the most effective activities (Sustainability Network and Ontario Ministry of the Environment 2000, IJC 1998).

Lesson Four—Ownership

The formulators of the plan are the owners of the plan. Ownership means that when agreeing to the plan, each member overtly recognizes and takes responsibility for the resource implications for its stakeholder group. Ownership results in pride in delivery, which sustains the process.

The community groups sought advice from the government team regarding technical and nontechnical options that could help restore beneficial uses. In the more successful RAP models, however, the RAP Team was not the decision-making body, but was a resource offered to the community. The provincial and federal government representatives did not have a formal voice in selecting the preferred plan. The agencies acted as advisors, informing the PAC of the technical feasibility, scientific certainty, and policy implications of their recommended plans. The PAC was advised, as well, that the governments might reject any plans that were inconsistent with government policy and the goals of the GLWQA. At such junctures the PAC often reconsidered their recommendations in light of new information and adjusted their approach.

Lesson Five—Respect

Trust and respect derived from a common purpose and reliability, strengthened the group's credibility and ability to solve challenges as a team.

The assignment of a RAP Coordinator by governments to assist the community in restoring their harbour brings with it complications as well as benefits. First, local leaders are correct to be skeptical that a government representative has arrived from outside the community with the good intent of helping them correct historic and ongoing problems. Trust is earned not granted. If the process is to work, confidence in one another needs to be fostered. Honest effort is more important than bureaucratic rhetoric. Within months, all members of the PAC and RAP team recognized that they truly did share the common goal of restoring the aquatic ecosystem. This was unifying, and trust, respect and honesty evolved among the participants.

Further, credibility was gained by making realistic and achievable commitments to the community at large, particularly at the level of local governments. Municipalities are extremely important partners in helping to implement a RAP. They make decisions that can protect environmental quality and preserve sensitive and valuable natural features. Land-use planning, investments in infrastructure, water and energy conservation, sewer use bylaws and other tools and practices proffer appreciable opportunities to advance the mission of the RAP. The agency staff and PAC forged a solid relationship with councillors and senior staff, and linked the RAP's planning needs with those of the municipality, sharing data and information (see Sustainability Network and Ministry of the Environment 2000).

Lesson Six—Incentives

The incentives for achieving the shared goal can differ among the participants and must be respected. It is not important that different stakeholders extracted different benefits by meeting the shared goals.

As the RAP process advanced, it was clear that participants wanted the ecosystem restored for different reasons. Some wanted recreational boating or fishing opportunities, others were concerned over the ability to eat the fish or swim in the water. Some wanted better birding or passive recreation, others looked at improved property values or growth of the tourism industry.

> Southeast Michigan, which includes four AOCs, has done an excellent job of showing leadership regarding wastewater infrastructure needs. In Wayne County, Michigan, the infrastructure cost (excluding inflation and interest) of combined sewer overflow control from 2001 until 2030 was estimated from a low value of $1.8 billion to a high value of $2.7 billion U.S. Sanitary sewer overflow remediation from the same time period was estimated with a low value of $40 million, and a high value of $431 million. In addressing these infrastructure needs, Wayne County is faced with estimated upgrade costs ranging from a low value of $1.8 billion, a mid-range value of $2.4 billion, and a high-end value of $3.2 billion. The entire Southeast Michigan needs study is an admirable representation of multijurisdictional coordination of multi-billion dollar efforts to mitigate impacts of aging and outdated infrastructure. It demonstrates the necessity of municipal participation in the RAP clean-up efforts. http://www.semcog.org/products/pdfs/sewerneeds2.pdf.

The important point was to be able to return to the crux of agreement, that the ecosystem needed to be restored, and that there was a shared vision regardless of individual motives. This allowed the participants to appeal directly to the interests of a diverse cross section of citizens and opinion leaders in ways that compelled others to take action (see Bonk et al. 1998).

Lesson Seven—Quality of Life

Recognition that the local economy and quality of life is inextricably bound to environmental excellence provides an impenetrable shield to the current economic-environment dialectic.

Lesson Eight—Measure Success

As partners from numerous sectors see the RAP participants are making incremental progress in restoring beneficial uses, more volunteers ask to participate and join in the successes.

In fact, the economic benefits of environmental health enabled the PAC to forge numerous partnerships with the business community in implementing a myriad of projects and programs. In the Case of Collingwood Harbour the PAC organized annual campaigns around habitat rehabilitation and public outreach with the support of service clubs, schools, the municipality, and donations from business of food, equipment, and advertising. The signature outreach project, ENVIRONPARK, was possible through a consortium of over 30 different local partners, with additional funding from the provincial and federal governments. Even the sediment clean-up project brought forward volunteers and local businesses to launch the initial clean-up of the shipyard boat slips, so that dredging could commence (see Krantzberg and Houghton 1996).

The Severn Sound Environmental Association (Association) is a partnership founded in 1997 with representation from 2 towns, 6 townships, Environment Canada, and the Friends of Wye Marsh, Inc. Its goals include restoring water quality in Severn Sound so that it can be removed from the list of AOCs and to assist in the transition of the local remedial action plan effort to a locally sustained environmental office. The Association seeks to become a model "Sustainable Community". The Association brings together the community and the resources of the federal, provincial and municipal governments and has implemented hundreds of water quality improvement projects. The Association has proven to be a very effective organization that has built the local community capacity to sustain restoration and economic vitality, in partnership with governments.
http://www.severnsound.ca/

The Maumee River RAP has benefited from active support of personnel of the U.S. Environmental Protection Agency, the Ohio Environmental Protection Agency, the Toledo Metropolitan Council of Governments as well as an active community committee. A feature of the Maumee River RAP is to document success and broaden partnerships. Recently the RAP produced a 263 page summary of activities and accomplishments that covers the period of 1991 to 2001. Measuring and celebrating progress as the Maumee RAP participants do, is fundamental to sustaining momentum for RAP implementation and attracting new volunteers.
http://www.maumeerap.org/

Lesson Nine—Leadership and Focus

Find a strong leader and stay focused on the task at hand. This allows for steady progress in the selection of remedial measures, their implementation, and the recovery of the ecosystem.

A fundamental ingredient in the successful RAP mix was the strong and directed leadership of the Chair who insured the mission of the RAP did not deviate from the need to restore the harbour ecosystem.

Concluding Remarks

The international Joint Commission's Water Quality Board, in 1996, concluded that RAPs are on the cutting edge of community-based and ecosystem based management processes (IJC 1996). RAP implementation and progress towards watershed management can continue to thrive with strong local leadership, despite reductions in some state, provincial and federal programs. Governments should be viewed as facilitators of RAPs and partnership builders, and must continue to provide resources and technical assistance to leverage local and private sector resources. Participation of the appropriate actors, development of mutually agreed upon decision-making processes, development of common objectives, dispute resolution, political support, public participation and funding are all central prerequisites to achieving the ecosystem approach (Mackenzie 1996), an approach inherent in successful RAPs.

RAPs can continue to be a source of pride and optimism, but this requires that the process be improved to make it more visible, inclusive, and institutionally integrated (Grima 1997). It also would benefit from a greater emphasis on measuring, celebrating and marketing successes, and building the local capacity to sustain progress.

The 1992 United Nations Conference on Environment and Development identified capacity-building in Agenda 21 as one of the essential means to implement sustainable development. Capacity-building means enhancing the ability of a community, region, or country to identify and reach agreement on problems, develop policies and programs to address them, and mobilize appropriate resources to fulfil the policies and programs (Hartig et al. 1995). Successful RAPs employed a combination of human, scientific, technological, organizational, institutional and resource capabilities to generate and sustain the capacity for the changes required to solve local environmental problems.

As observed by Hartig and Law (1994), RAPs require cooperative learning that involves stakeholders working in teams to accomplish a common goal under conditions that involve positive interdependence (all stakeholders cooperate to complete a task) and individual and group accountability (each stakeholder is accountable for the final outcome). For RAPs to be successful, they must:

- be clean-up- and prevention-driven, and not document-driven;
- make existing programs and statutes work;
- cut through bureaucracy;
- elevate the priority of local issues;
- ensure strong community-based planning processes;

- streamline the critical path to use restoration; and
- be an affirming process.

RAPs are an unprecedented collaboration of international significant worth continuing by the parties, the jurisdictions and the public (Krantzberg 1997). The passion and dedication of communities involved in implementing RAPs need ongoing nurturing. Solidarity does emerge, and potential adversaries become allies united by a vision of a shared inspiration to enhance and protect the magnificence that is the Great Lakes.

Community groups that have enlisted local champions can marshal incentives for others to join the mission. Creative, innovative partnerships and institutional arrangements are needed to stimulate and sustain advances in the clean-up of the AOCs, to control contaminant inputs, restore riparian vegetation, rehabilitate coastal wetlands, remediate contaminated sediment, raise public awareness of individuals' responsibilities, unite government with nongovernment leaders, and make the Lakes great.

References

Bonk, K., H. Griggs and E. Tynes. 1998. Strategic Communications for Nonprofits. Jossey Bass Publishers, San Francisco.

Canada-Ontario. 1999. Third Report of Progress Under the Canada-Ontario Agreement Respecting the Great Lakes Basin Ecosystem 1997–1999. http://www.on.ec.gc.ca/coa/third-progress-report.

Collingwood Harbour RAP. 1992. Stage 2 Document. Prepared by the Ontario Ministry of Environment and Energy, Toronto, Ontario; Environment Canada, Ministry of Natural Resources, and the Collingwood Harbour Public Advisory Committee. ISBN # 0-7778-0162-0.

Collingwood Harbour RAP. 1994. Stage 3 Document. Right On Target. Prepared by the Ontario Ministry of Environment and Energy, Toronto, Ontario; Environment Canada, Ministry of Natural Resources, and the Collingwood Harbour Public Advisory Committee.

Costanza, G. 1992. Ecological economics of sustainability: investing in natural capital. pp. 106–118. *In*: R. Goodland, H.E. Daly and S.E. Serafy (eds.). Population, Technology and Lifestyle. Island Press, Washington DC.

Davidson, P.W., G.J. Myers, C. Cox, C.F. Shamlaye, D.O. Marsh, M.A. Tanners, M. Berlin, J. Sloane-Reeves, E. Cernichiari and O. Chloisy. 1995. Longitudinal neurodevelopmental study of Seychellois children following in utero exposure to methylmercury from maternal fish ingestion: outcomes at 19 and 29 months. Neurotoxicology 16: 677–688.

Environment Canada. 1997. Great Lakes 2000 Cleanup Fund Project Summaries Report. http://www.on.ec.gc.ca/glimr/data/cleanup-project-summaries/intro.html ISBN 0B662-26232-8.

Grima, A.P.L. 1997. Ten Years of RAPs: Reflections and suggestions. Journal of Great Lakes Research 23: 232–233.

Gurtner-Zimmermann, A. 1995. A mid-term review of Remedial Action Plans: Difficulties with translating comprehensive planning into comprehensive actions. Journal of Great Lakes Research 21: 234–247.

Gurtner-Zimmermann, A. 1996. Analysis of Lower Green Bay and Fox River, Collingwood Harbour, Spanish Harbour, and the Metro Toronto and Region Remedial Action Plan (RAP) Processes. Environmental Management 20: 449–459.

Harris, C.C., E.A. Nielsen, W.J. McLaughlin and D.R. Becker. 2003. Community-based social impact assessment: The case of salmon-recovery on the lower Snake River. Impact Assessment and Project Appraisal 21: 109–118.

Hartig, J.H. and M.A. Zarull (eds.). 1992. Under RAPs. Toward Grassroots Ecological Democracy in the Great Lakes Basin. University of Michigan Press, Ann Arbor, Michigan.

Hartig, J.H. and N.L. Law. 1994. Institutional frameworks to direct the development and implementation of Great Lakes remedial action plans. Environmental Management 18: 855–864.

Hartig, J.H., N.L. Law, D. Epstein, K. Fuller, J. Letterhous and G. Krantzberg. 1995. Capacity-building for restoring degraded areas in the Great lakes. International Journal of Sustainable Development & World Ecology 2: 1–10.

International Joint Commission. 1978. Great Lakes Research Advisory Board. The Ecosystem approach. Special Report to the International Joint Commission. Windsor, Ontario.

International Joint Commission. 1996. Position statement on the future of Great Lake Remedial Action Plans. Report of the Water Quality Board. Windsor, Ontario, Canada.

International Joint Commission. 1997. Overcoming Obstacles to Sediment Remediation. Report of the Sediment Priority Advisory Committee (SedPAC) to Water Quality Board.

International Joint Commission. 1998. If You Don't Measure It, You Won't Manage It! Measuring and Celebrating Incremental Progress in Restoring and Maintaining the Great Lakes. Report of the Water Quality Board.

International Joint Commission. 2002. Eleventh Biennial Report on Great Lakes Water Quality Ottawa, Washington, Windsor.

International Joint Commission. 2003. The Great Lakes Areas of Concern Report. Windsor, Ottawa, Washington.

Jacobson, J.L. and S.W. Jacobson. 1996. Intellectual impairment in children exposed to polychlorinated biphenyls in utero. New England journal of Medicine 335: 783–789.

Krantzberg, G. 1997. International association for Great Lakes research position statement on Remedial Action Plans. Journal of Great Lakes Research 23: 221–224.

Krantzberg, G. and E. Houghton. 1996. The Remedial Action Plan that lead to the cleanup and delisting of Collingwood Harbour as an Area of Concern. Journal of Great Lakes Research 22: 469–483.

Krantzberg, G., H. Ali and J. Barnes. 1999. What progress has been made in the RAP program after ten years of effort? pp. 1–13. *In*: T. Murphy and M. Munawar (eds.). Aquatic Restoration in Canada, Ecovision World Monograph Series. Backhuys Publishers, The Netherlands.

Lee, B.J., H.A. Regier and D.J. Rapport. Ten ecosystem approaches to the planning and management of the Great Lakes. Journal of Great Lakes Research 8: 505–519.

Lonky, E., J. Reihman, T. Darvill, J. Mather, Sr. and H. Daly. 1996. Neonatal behavioral assessment scale performance in humans influenced by maternal consumption of environmentally contaminated Lake Ontario fish. Journal of Great Lakes Research 22: 198–212.

Mackenzie, S.H. 1996. Integrated Resource Planning and Management: The Ecosystem Approach in the Great Lakes Basin. Island Press, Washington.

Managing Shared Waters. 2002. http://www.pollutionprobe.org/managing.shared.waters/.

Ontario Ministry of Environment. 2001. Guide to Eating Sport Fish 2001–2002. Queen = Printer. ISSN 0826-9653.

Samy, M., H. Snow and H. Bryan. 2003. Integrating social impact assessment with research: The case of methylmercury in fish in the Moblie-Alabama River Basin. Impact Assessment and Project Appraisal 21: 133–140.

Schwartz, P.M., S.W. Jacobson, G. Fein, J.L. Jacobson and H.A. Price. 1983. Lake Michigan fish consumption as a source of polychlorinated biphenyls in human cord serum, maternal serum, and milk. American Journal of Public Health 73: 293–296.

Sproule-Jones, M. 2002. The Restoration of the Great Lakes. University of British Columbia Press, Vancouver.

Sustainability Network and Ministry of the Environment. 2000. The Road to Delisting: Addressing RAP Challenges. http://www.sustain.web.net/resource/pdf/rapreportPDF.

Town of Collingwood. 2001. http://www.town.collingwood.on.ca/ProfileTOC.cfm.

United States and Canada. 1972. Great Lakes Water Quality Agreement.

United States and Canada. 1987. Revised Great Lakes Water Quality Agreement of 1978 As Amended by Protocol Signed November 18, 1987. Consolidated by the international Joint Commission.

Example of a Green Building

Algonquin Centre for Construction Excellence (ACCE) –Algonquin College (Cowan 2013, Guly 2012)

Acting as a new gateway for the campus, this facility explores innovative new environments in skilled trades education.

Location: Ottawa, Ontario
Size: 18,460 m² or 198,708 ft²
Certification: LEED Canada Platinum NC

Green Technologies and Features

Materials

- A 5-storey biowall of live tropical plants that produce oxygen that is drawn into the ventilation system to improve air quality.
- Building materials and finishes have been selected to minimize VOC exposure to the occupants.
- The building envelope has triple glazed R-8 windows and an R-50 roof vegetated system.

Water Conservation

- Storm Water Capture System to capture rainwater from the roof and is used to flush urinals and toilets and irrigate the vegetated roof. This system controls the discharge storm water flow back into an adjacent local creek.
- The 4,000 m² green roof offsets the sunlight and lowers surrounding air temperatures. As well, the vegetated sedums absorb and control storm water runoff which reduces the strain on the City of Ottawa's storm sewer system.

Energy

- A solar domestic hot water system that transfers heat from heat exchangers to the domestic water system.
- A solar wall on the building's west side pre-heats the intake air for HVAC system.
- An integrated control system for lighting, daylight and shading are in place.
- Building orientation and siting combine with solar shading to minimize heat gain.

Ventilation

- A hybrid hydronic heat pump system is distributed throughout the building.
- Ninety-five percent efficient heat-recovery wheels have helped achieved a 68 percent betterment over the Model National Energy Code for Buildings.

Other Innovations and Technology

- Exposed building and environmental systems aid student awareness and understanding of the building infrastructure.
- Sensors provide real-time and historical building diagnostics allowing students to monitor outputs such as temperature, humidity and air quality.
- Several hundred trees and over twenty varieties of groundcover have increased the local bio-diversity by 300 percent.

Picture and Description, Courtesy: Tony Cupido

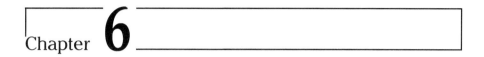

Chapter 6

Sustainable Brownfields Redevelopment: Case Study, Collingwood Harbour, Georgian Bay, Ontario

Introduction

In 1972, the Governments of Canada and United States created the Great Lakes Water Quality Agreement which was signed by Prime Minister Trudeau of Canada and President Nixon of the United States (United States and Canada 1972). In November 22, 1978, a revision to the 1972 Agreement provided new programs directed predominantly towards "virtually eliminating inputs of persistent toxic substances to the Great Lakes" (United States and Canada 1978). The governments adopted the revolutionary concept of applying an ecosystem approach to enhancing and maintaining the health of the Great Lakes. By November 1987, dissatisfied with progress on Great Lakes revitalization, the Governments signed a Protocol to the 1978 GLWQA that included Remedial Action Plans (RAPs) under Annex 2 of the Agreement. The United States and Canada revision in 1987 of the GLWQA commits the countries, in collaboration with the states and provinces, to develop RAPs at Areas of Concern (42 in total at that time, Fig. 6.1) (United States and Canada 1987). A RAP is an approach that brings governments and concerned citizens together to restore and protect 'beneficial uses' (14 of which are specified in the Agreement). Importantly, the governments are to work and with local communities to jointly develop and implement the RAPs. Krantzberg (2003) and Gurtner-Zimmermann (1995), discuss the process for developing and implementing RAPs in greater detail.

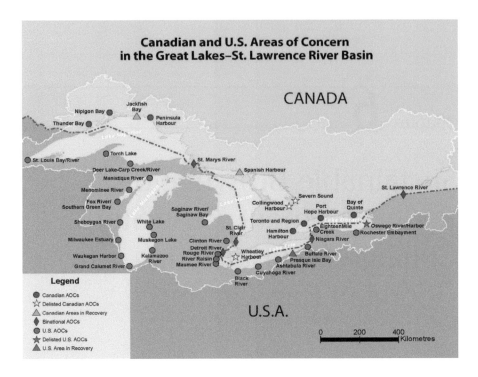

Figure 6.1. Location of the Great Lakes Areas of Concern. *Source*: http://www.epa.gov/glnpo/aoc/.

The reliance of the economy on a healthy Great Lakes Basin Ecosystem is unequivocal. Degradation of environmental quality directly downgrades the potency of the region's economy and quality of life, and the sustainability of cities and towns. This chapter investigates whether a community-based, participatory RAP processes can generate capacity to ensure sustainable redevelopment of a particularly degraded portion of the Town of Collingwood's waterfront, a brownfield site.

Different definitions in both Europe and the US describe brownfield sites as abandoned or underused properties, for which intervention is required to ensure beneficial reuse because of the real or suspected presence of hazardous substances, pollutants or contaminants. The health and economic threats of brownfields as well as the challenges and potential of their reuse are recognized world-wide and international literature describes concerns related to brownfields (Schädler et al. 2011).

With up to 1,000,000 brownfields in the US alone, the magnitude of the benefits from redevelopment can be significant. These may include the revitalization of blighted areas, promotion of 'smart growth' development, reduction of development pressure on greenfields, mitigation of issues relating to environmental justice, reduction of risk to public health, and economic growth (Wedding and Crawford-Brown 2007). With a view to a more healthy and sustainable water/land interface, the approach used in Collingwood demonstrates possibilities of success.

Sustainability herein is defined as the balanced relationship between dynamic human economic systems and the dynamic, but generally slower-changing ecological systems in which civil society can develop, but within bounds such that human activities do not destroy the diversity, complexity, and function of the ecological life-support system (Costanza 1992). To achieve sustainability, ecology, economics and society must be integrated.

Case Study: Collingwood Harbour

As shown in Fig. 6.2, Collingwood Harbour is positioned on south shore of Nottawasaga Bay, the southern extension of Lake Huron's Georgian Bay. The town of Collingwood sits south of its harbour with a population of approximately 21,500. Collingwood was the railhead of Ontario during the mid to late 1800s, and its harbour was the transshipment point for goods destined to western Canada. In 1883, the Collingwood Shipyards, then known as Collingwood Dry Dock Shipbuilding and Foundry Company Limited, opened. The Shipyards became one of the principal industries in the town. Habitat and wetland loss, shoreline hardening, and contaminated sediment were a result of this industrial legacy (Krantzberg and Houghton 1996). With

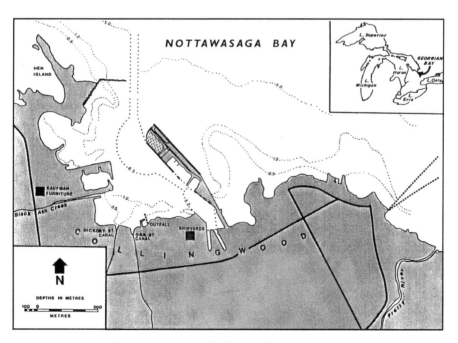

Figure 6.2. Location of Collingwood Harbour, Ontario.

the closure of the shipyards in 1986, the largest privately owned parcel of land in the Town remained an idle brownfield.

In its final publication to Town Council, the RAP's 'Sustainability Subcommittee', chaired by the Town Planner, wrote that "development (at the shipyards) is acceptable so long as certain conditions are met to ensure water quality and the promotion of fish habitat" (Collingwood Harbour RAP 1994).

Finally, in 2004, plans were being finalized for the development of a mixed residential commercial project occupying the shipyard's property, approximately 16.6 hectares along the Harbour Waterfront. The purpose of this research was to decipher whether the principles of the RAP were effective in enabling sustainable growth along the waterfront and within the Town 10 years after RAP implementation was completed and the harbour delisted as an Area of Concern.

Sustainable Communities in Relation to Collingwood

The Collingwood Harbour RAP practitioners[1] acknowledged and deployed an ecosystem approach using the man-in-system definition (IJC 1978). Practitioners conceived of the ecosystem as an interacting lattice of water, air, land and living organisms including man.

With human inclusion in the ecosystem approach, societal values were overtly expressed. A predictor that the RAP would leave a sustainable legacy is the civic recognition among the town participants that the local economy and quality of life is inextricably bound to environmental excellence. This unequivocal reality effectively defused what can become a divisive economic-environment dialectic inconsistent with the ecosystem approach to resource revitalization.

Sustainable cities are vibrant, harmonious and lasting (Sustainable Cities 2004). Like Collingwood, they are a pleasure to live in. According to Florida (2000) environmental excellence exceeds other factors including housing costs, climate, government services and public safety in the selection of places to live.

Sustainable communities have also been defined as places that have made it their business to remain robust over the long term. "Sustainable communities have a strong sense of place. They have a vision that is embraced and actively promoted by all of the key sectors of society, including businesses, environmentalists, civic associations, government agencies, and more. They are places that build on their assets. These communities value healthy ecosystems and actively seek to retain and enhance a locally based economy... Public debate in these communities is engaging, inclusive, and constructive" (Institute for Sustainable Communities 2004).

---

[1] Provincial and federal government experts, community representatives from a cross section of sectors including local government, education, environment agriculture, industry, tourism and recreation, land owners, conservation authority, human health, citizens at large.

The RAP was effective and efficient in resolving the consequences of historic misuse (Krantzberg and Houghton 1996). This is in part a function of the inclusiveness of stakeholder representation and goal setting. Inclusivity lends legitimacy, stimulates accountability, and can galvanize potentially adversarial stakeholders.

Boyd (2004) refers to the emergence of sustainable communities that convey ambitious new environmental, economic, and social agendas. In order to advance a prosperous, just and sustainable future, Boyd lists nine critical challenges. We emphasize three here:

- PROTECTING AND CONSERVING WATER: Recognizing and respecting the value of water in our laws, policies, and actions,
- CONSERVING, PROTECTING AND RESTORING CANADIAN NATURE: Taking effective steps to stop the decline of biodiversity and revive the health of ecosystems, and
- BUILDING SUSTAINABLE CITIES: Avoiding urban sprawl in order to protect agricultural land and wild places, and improve our quality of life.

These challenges are important determinants of the sustainability of Collingwood's brownfield redevelopment.

Civic governance works well when there is a civic community marked by an active, public-spirited citizenry, by egalitarian political relations, by a social fabric of trust and cooperation (Putnam 1993). Such qualities were cultivated during the development and implementation of the Collingwood Harbour RAP. The social fabric of the community encouraged cooperation.

A strong sense of place-based consciousness sways individuals to act cooperatively. It provides the context for cooperative action according to ethics despite economic and immediate, need-meeting motivations which encourage one to do otherwise (Moore 1994). Civic consciousness supports sustainability.

A strong commitment from the local government to be inclusive means that involvement of the community needs to be transparent and expansive. Multi-stakeholder and citizen engagement can generate local ownership of a community's policies and programs with the community. The sincere integration of community members as partners in decision-making stimulates successful community involvement. These are characteristics demonstrated within Collingwood and among the RAP practitioners (Krantzberg 2003).

The 10 Aalborg Commitments help local governments set qualitative and quantitative targets to implement the urban sustainability principles of the Aalborg Charter (Aalborg 2004). The Aalborg Commitments address 10 themes that are instructive to the North American context at the local level for ensuring sustainability. Some of these are:

- Governance—Local governments increase citizens' participation and cooperation with all spheres of governance in their efforts to become more sustainable.

- Urban management—Local governments formulate, implement and evaluate management schemes aimed at improving urban sustainability.
- Natural common goods—Local governments preserve natural common goods.
- Planning and design—Urban planning is vital in addressing environmental, social, economic, and health issues.
- Local action for health—Local governments have a duty to protect the health of their citizens.
- Sustainable local economy—Local governments are committed to creating a vibrant local economy that promotes employment without damaging the environment.

Discussion

"Vision 2020 Report No. 6—The Waterfront" (Town of Collingwood 2002) was compiled by a community committee. The report recommends to Town Council:

*"that a proactive **Ecosystem approach** be taken when considerations are made. Within an ecosystem approach everything is connected to everything else. These links are air, soil, water, wildlife, land uses, communities, economic activities and much more. If we do this we will understand how the parts affect one another and we will understand the complexities of the whole.*

*As a result of our ecosystem approach we can identify ways in which human activities can be reintegrated into the ecological process to ensure efficient use of resources, reduce waste and pollution, etc. It is also incumbent upon an ecosystem approach to hold that the economy, social issues, and the environment are inter-related. An ecosystem approach would make the most of the quality of Collingwood's waterfront area and ensure our responsibilities to future generations. A vibrant, healthy waterfront is what we need to accomplish the above. The waterfront is our **crowning glory.** It represents best the future of our community. It is what the citizenry aspires for."*

Earlier it was noted that the Shipyards remained an undeveloped brownfield adjacent to Collingwood Harbour at the time of delisting in 1994. The final development design of this critical piece of waterfront was revealed in 2003/04 (Fig. 6.3). The Lands affected by the proposed Official Plan Amendment (the former C.S.L. shipyard property) are on the southern shore of the Collingwood Harbour, at the northern end of Hurontario Street, which is the main street connecting the downtown core to the harbour and Georgian Bay. The site is approximately 16.6 hectares (41.0 acres) in area.

Citing from the development plan:

"Something monumental is happening to Collingwood's downtown. The entire shoreline is about to be transformed...The pleasure of downtown conveniences, the beauty of Georgian Bay's breathtaking water and mountain views, and the chance to discover four-season living at your doorstep makes

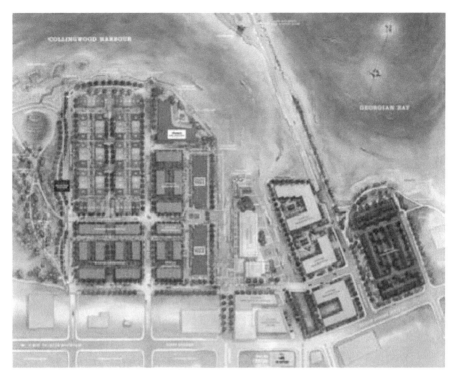

Figure 6.3. The Shipyards proposed development plan.

this community the only one of its kind. Be drawn to the water's edge. There is so much to experience at The Shipyards. Embrace the best of all-inclusive waterfront living at The Shipyards (Fram Building Group and Slokker Canada Corp 2004)."

Included in the development plans is an open space situated on the west side of the property and will be accessible to the public, a Natural Common Good (Fig. 6.2). This feature was recommended by the RAP's 'Sustainability Subcommittee' in 1994. Urban planning and design reflected good governance in adapting to the desires of the citizenry. Local action for health is being ensured, in that before any of the construction takes place, soil remediation of the brownfield will be completed. Importantly, a wetland feature, providing fish and wildlife habitats is included in the proposal, and is in keeping with the 'Sustainability Subcommittee' and 1994 RAP document (Collingwood Harbour RAP 1994). Waterfront trails, a public plaza and a recreational facility for public use, are also included in the plans, and are consistent with the purpose of sustainable towns, to support the environmental, economic, and social fabric of the community.

Years after the delisting of the Harbour, bringing people back to the revitalized waterfront has been a Town priority. Harbourlands Park was created in 2000 and is one of the most beautiful areas in the community. Residents and visitors alike are enjoying the rugged beauty of a once active shipping/grain storage area. The backdrop of the Collingwood Terminals with its huge white columns rises up from the once wasteland 'spit area', now a series of beautifully landscaped walkways and gardens with a history of the area on massive granite plinths. Harbourlands Park offers the ever-changing grandeur and scenic beauty of Georgian Bay for the many people who drive or walk to the Park. There are benches for reflective moments or to watch the quiet beauty of sailboats filling their sails as they make their way out of the historic Collingwood Harbour (Krantzberg 2006, Town of Collingwood 2004).

Conclusions

According to the IJC (1996), the Remedial Action Plan process is breaking ground in community-based and ecosystem based management processes. In 2008, some might argue that the ground breaking in many locations has stalled. Where this is the case, from the Collingwood experience, it is apparent that governments need to view themselves as a mechanism that nurtures community capacity. Participation of the appropriate actors, development of mutually agreed upon decision-making processes, common objectives, political support, public participation and funding are all central prerequisites to achieving a sustainable community, and are central to the philosophy that is behind successful RAP programs (Krantzberg 2003).

The Collingwood Harbour RAP employed a combination of human, scientific, technological, organizational, institutional, and resource capabilities to generate and sustain the capacity for the changes required to solve the Harbour's environmental problems. As defined by Hartig et al. (1995) capacity-building enhances the ability of a community to identify and achieve consensus on problems, develop policies and programs to address them, and marshal appropriate resources to carry out the policies and programs. Further, democratic dialogue and participatory decision making enabled consensus and ownership of the RAP and its legacy. This is evidenced beyond the RAP by the commitment of the politicians and citizens to ensure the sustainability of the Town's economy, environment, and social fabric. Cumulatively, the consequence encompassed a community fiercely protective of its excellence that embedded sustainable design in the approved redevelopment plan of the brownfield property at the Shipyards.

Acknowledgements

We thank my most esteemed associate, Douglas Markoff for providing the stimulus to document this remarkable experience. We continued thanks go to Ed Houghton, the Collingwood Harbour Public Advisory Committee Chairman, unwavering leadership of the mayors of the Town of Collingwood, and the proud people of this extraordinary Town.

References

Aalborg. 2004. http://www.iclei.org/home/documents/press_release_11june2004.doc.

Boyd, D.R. 2004. Sustainability within a Generation. A New Vision for Canada. David Suzuki Foundation, Canada.

Collingwood. 2004. http://www.town.collingwood.on.ca/.

Collingwood Harbour RAP. 1994. Planning for the Future, A report to Collingwood Town Council. Coastline Development and Sustainability Subcommittee, Collingwood, Ontario.

Costanza, G. 1992. Ecological economics of sustainability: investing in natural capital. pp. 106–118. *In*: R. Goodland, H.E. Daly and S.E. Serafy (eds.). Population, Technology and Lifestyle. Island Press, Washington DC.

Florida, R. 2000. Competing in the age of talent: quality of place and the new economy. http://www.heinz.cmu.edu/~florida.

Gurtner-Zimmermann, A. 1995. A mid-term review of Remedial Action Plans: Difficulties with translating comprehensive planning into comprehensive actions. Journal of Great Lakes Research 21: 234–247.

Hartig, J.H., N.L. Law, D. Epstein, K. Fuller, J. Letterhous and G. Krantzberg. 1995. Capacity-building for restoring degraded areas in the Great lakes. International Journal on Sustainable Development and World Ecology 2: 1–10.

International Joint Commission. 1978. Great Lakes Research Advisory Board. The Ecosystem Approach. Special Report to the International Joint Commission. Windsor, Ontario.

International Joint Commission. 1996. Position Statement on the Future of Great Lake Remedial Action Plans. Report of the Water Quality Board. Windsor, Ontario, Canada.

International Joint Commission. 2003. The Great Lakes Areas of Concern Report. Ottawa, Washington, Windsor.

Krantzberg, G. 1997. International association for great lakes research position statement on remedial action plans. Journal of Great Lakes Research 23: 221–224.

Krantzberg, G. 2003. Keeping remedial action plans on target: Lessons learned from Collingwood Harbour. Journal of Great Lakes Research 29: 641–651.

Krantzberg, G. 2006. Sustaining the gains made in ecological restoration: Case study Collingwood Harbour, Ontario. Environment, Development and Sustainability 8: 413–424.

Krantzberg, G. and E. Houghton. 1996. The remedial action plan that lead to the cleanup and delisting of Collingwood Harbour as an Area of Concern. Journal of Great Lakes Research 22: 469–483.

Moore, J.L. 1994. What's Stopping Sustainability? http://www.newcity.ca/Pages/mooreindex.html.

Putnam, Robert, Robert Leonardi and Raffaell Y. Nanetti. 1993. Making Democracy Work: Civic Traditions in Modern Italy. Princeton University Press, Princeton, N.J.

Schädler, S., M. Morio, S. Bartke, R. Rohr-Zänker and M. Finkel. 2011. Designing sustainable and economically attractive brownfield revitalization options using an integrated assessment model. Journal of Environmental Management 92: 827–837.

Sustainable Cities, 2004. http://www.sustainable- cities.org.uk/institute/index.html.

Town of Collingwood. 2002. Vision 2020 Report No. 6—The Waterfront http://www.town.collingwood.on.ca/living_news.cfm?category=25.

Town of Collingwood. 2004. http://www.collingwood.ca/visiting_tourism.cfm?action=list&type=1&topic=1.

United States and Canada. 1972. Great Lakes Water Quality Agreement.

United States and Canada. 1987. Revised Great Lakes Water Quality Agreement of 1978 As Amended by Protocol Signed November 18, 1987. Consolidated by the international Joint Commission.

Wedding, G.C. and D. Crawford-Brown. 2007. Measuring site-level success in brownfield redevelopments: A focus on sustainability and green building. Journal of Environmental Management 85: 483–495.

Example of a Green Building

Enermodal Engineering's Headquarters (Douglas, J. 2011)

This facility is the best example of a premier energy efficient and sustainable building in Canada. It has optimized the opportunity provided for LEED certification and made best use of fundamental engineering principles in building planning, design, construction and operations.

Location: Kitchener, Ontario
Size: 2,045 m² or 22,000 ft²
Certification: LEED Canada Platinum NC, CI and EB: O&M

Green Technologies and Features

Energy

- Exceptional energy use intensity target of 69 kilowatt-hours per square metre (kWh/m^2) when compared to the national average of 396 kWh/m^2 for a similar building.
- An Energy Star score of 100, which indicates that this facility has a lower energy use than 100% of peer or like buildings.
- An east-west orientation for the building.
- Triple-glazing with fibreglass frames and insulated concrete form envelope for higher R rating.
- Solar gain and glare are controlled by automated exterior shades.

Ventilation

- Incoming outdoor air is moderated by the earth via concrete earth tubes and then into the building. This approach has significant energy savings.
- Separate rooftop air-source heat pumps for each occupied floor which are connected to 60 small fan coil units thus distributing the heating and cooling throughout each floor.
- Thermostats and occupancy sensors are connected to the building automations system through wireless infrared signal to improve control and energy consumption.
- Heat from the IT server rooms is reused to preheat domestic hot water demand.

Operational/Occupational Approaches

- A Green Team was established to help oversee the development, implementation and documentation of sustainable Policies, inclusive of procurement of office products, implementation of green housekeeping, "green" landscaping techniques such as no power tools, no pesticides and no irrigation.
- Establishing employee garden plots for staff to grow vegetables and fruits.
- Reduced transportation-related carbon emissions to improve EB: O&M point capture by utilizing video conferencing for their Canada—wide offices.
- Procurement of a hybrid vehicle for their Kitchener based employees.
- Utilization of an in-house-developed software program that allows employees to log their daily commute in an effort to monitor and reduce their commuting footprint.
- Full reimbursement for employee purchases of compost bins, rain barrels and water-efficient shower heads.
- An electronic touch-screen kiosk in the main lobby that provides real-time data on energy performance for employees and visitors.

Chapter 7

Monitoring and Measuring Urban Regeneration[1]

Introduction

Urban regeneration projects add to a vision of renewing old, decaying parts of the cities, typically downtowns but not solely. Globally there have been many cases of successful urban regeneration, some examples include in North America (Chicago, Portland), Europe (particularly in the U.K., La Defence in France, Bilbao in Spain, South Pact, Rotterdam in the Netherlands, Oath Basin, Porta Palazzo, Turin in Italy), Asia (Japan, Korea, Vietnam and Kazakhstan) (Park et al. 2008, Colantonio and Dixon 2009).

In the 1980s, regeneration projects in Europe focused on the physical and economic renewal of the decayed and degraded inner city areas. However, since the 1990s, this has been replaced by a more holistic and integrated approach which links stimulation of economic activities and environmental improvements with social and cultural elements. In this approach, concepts of 'community' and 'neighborhood' have become a central focus of analysis (Colantonio and Dixon 2009). Thus the three pillars of urban regeneration are to improve economic, social and environmental conditions along with urban sustainability (Cubin et al. 2006). Regeneration projects are supposed to bring about changes to the local economy, neighborhoods and people. Hence it is important to see urban regeneration from a broader perspective and not just individual projects (i.e., the combined impact of several renewal projects should be seen in an integrated manner for its impact, consequences and unintended consequences). Urban regeneration projects include revitalizing space for businesses, residences, culture and leisure (Park et al. 2008).

---

[1] Based on the chapter on "MONITORING AND MEASURING URBAN REGENERATION" in Urban Regeneration: The Hamilton Brand (Dec 2013) Co-written with Gail Krantzberg and William Humber.

Urban Regeneration can help to improve urban life by redeveloping the old residential houses, high-rise buildings, businesses and by improving public facilities so that local communities are revitalized together with businesses in the area. Similar views are expressed by the Korean Urban Renaissance Center that believes that an old urban district can be revitalized by carrying out the repairs in a systematic fashion by introducing/inducing unique socio-cultural characteristics through a diversity of participants including regional industry and thus strengthening the overall economy (Park et al. 2008).

Large scale urban regeneration projects would involve (Park et al. 2008):

- large scale financial investment;
- dealing with complex social, political and legal uncertainties and challenges;
- satisfying public benefits with a requirement to meet higher standards of public well-being;
- crafting a quality landscape that integrates with existing surroundings.

It is important that a vision and strategy for a regeneration project be established, followed by the selection of the right set of indicators and how to monitor and analyze them. This chapter defines the terms such as monitoring, indicators, evaluation and result based management; discusses the importance of evaluation and monitoring in general followed by some specific examples of indicators used in the case of regeneration for governments and builders. This chapter also gives checklists that can be used while creating a monitoring and evaluation program.

Monitoring and Evaluation Principles

Some of the terms used, such as monitoring, evaluation and indicators, will be discussed here:

Monitoring

> *"Monitoring is the systematic and continuous assessment of the progress of a piece of work over time. It is a basic and universal management tool for identifying strengths and weaknesses in a project or programme. Its purpose is to help all the people involved, make appropriate and timely decisions that will improve the quality of the work"* (Gosling and Edwards 2003).

Monitoring can be defined as, "the systematic collection of data comprising key indicators of social, environmental and economic impacts" (Bankes and Thompson 1980, p. 10). Monitoring is an important part of cyclical program design, implementation and evaluation, because the information collected can be used to improve the decision-making process—either directly if used as a feedback tool for project management to improve policies (by looking at the success or failure of current policies, or to respond to new opportunities) or by indirectly informing the local community (or people) of the progress of project(s). There are multiple users of monitoring results: community, planners of the project, policy makers and developers (Carley 1986).

The most critical thing in monitoring is to decide what to measure, how to measure it and what to report. Some of the factors that need to be considered include answers to questions such as: what are the important social, cultural and economic issues facing local people? In the end issues are the main consideration and indicators should be selected in a way that they link to the key issues including critical social issues. Monitoring should be included from the planning stage and continue through the operational phase. For example in building construction, there should be indicators in the planning stage (see Fig. 7.1), construction stage (e.g., emissions during demolition or construction) and operation of the building (Carley 1986).

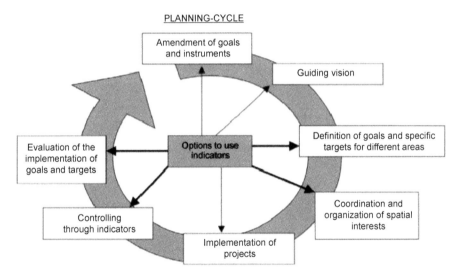

Figure 7.1. Indicators used in different phases of planning cycle. *Source*: Birkmann 2003.

In another case, indicators were chosen around key issues or concerns raised in a study for buildings/urban construction, which include (Häkkinen 2007):

1) Environmental impacts of building and use of buildings.
2) Health and comfort, quality of building.
3) Availability of housing and buildings for seniors (aging population) and disabled residents.
4) Readjustment for new business environment, full exploitation of new technological challenges.
5) Age of building stock, activities in refurbishment and renovation.

Some of the objectives of monitoring include: collection of data (systematic assessment of the progress); identifying strengths and weaknesses of the project for informed decision-making; feedback of success or failure (timely and continuous improvement) and local community involvement. Since monitoring gives updates of ongoing activities, it can also be used in evidence-based policy making. If done right, there can be multiple users of monitoring results:

Multiple users of monitoring results

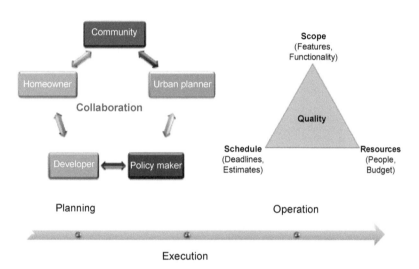

Evaluation

Evaluation can be defined as the comparison of actual project impacts against the agreed strategic plans (Umtha Strategy Planning and Development Consultancy).

There are three main conditions for evaluation: use a multi-dimensional perspective; have different 'instruments' of evaluation (including high level, intermediate or low level data tools); and applying expert judgment. Evaluation should be linked to: the characteristics of each program; ways in which different programs interact; and the ways in which normative frameworks are established (Breda-Vazquez et al. 2010).

Criteria that should be used (based on Spanish study, Breda-Vazquez et al. 2010, p. 216) in selecting an evaluation methodology include:

1) the overlapping of spatial scope;
2) local capacity to frame a particular issue, specifically their view of urban regeneration based on conceptualizations of the urban problems faced by the area in question;

3) the diversity of the objectives and of the agents involved in the decision making and implementation processes, reflecting the diversity and fragmentation of urban policies; and;
4) the existence of different stages of development: new initiatives, initiatives in progress, and completed initiatives.

Indicators

Historically, use of indicators can be traced back to the 1940s with the publication of monthly Economic Indicators to measure the buoyancy of the US economy which led to the development of indicators to measure social change in the mid-1960s (Wong 2003). The idea of using indicators then expanded to international organizations as well. Although, the indicators were initially used to measure economic and social progress, it was then expanded to include monitoring of environmental problems which led to the development of environmental, sustainability and quality-of-life indicators. Indicators are also used to assess and monitor the effectiveness of policy activities (evidence based policy), as well as for in performance measurement (Wong 2003).

An indicator can be defined as a way of measuring progress, some sort of quantum or determinable quantity for measuring a specific goal. It shows the direction of progress and improvement or the lack thereof. An indicator can also be defined as a policy-relevant variable that is specified and defined in such a way that it can be measured over a period of time and/or space (Astleithner et al. 2004, p. 4). The key feature of an indicator is that it can be measured, so it can be either a quantitative or a qualitative indicator. Indicators can be used to compare progress over a period of time, from one government to the other, from one project to the other or even within a project to measure performance of the project against its objectives, timeline and budget (Astleithner et al. 2004, p. 4).

Indicators can be used even to measure progress in everyday life. For example, if we set a goal to buy a house and start saving, the money in our account is an indicator of how close we are to achieving our goal. Another example, in a broader sense is, economic growth every year is an indicator of the economy, and over a period of time it shows the growth or downtrend in the economy (Grover 2001). Indicators are information signals on what is happening at one point and depict a trend over a given period of time.

Indicators serve three main functions: quantification, simplification and communications for both target setting and monitoring (Häkkinen 2007).

There are various sets of indicators for measuring sustainability, economy, urban regeneration, etc. and they can be usually classified under different headings depending on: level of application (project

"The identification of Sustainable Development Goals should not be data driven, but grounded in common values, relevant science, and a conceptual framework that represents key domains of sustainable development and inter linkages between the domains." Pinter, Jan 2013

level, state level, national level and global level); area of application (economic, environment and social etc.); way of representation; qualitative or quantitative (Reynolds and Cavanagh 2009).

According to the Chapter 40 of Agenda 21 "indicators of sustainable development need to be developed to provide solid bases for decision-making at all levels, and to contribute to a self-regulating sustainability of integrated environmental and development systems.". One example is, Global Reporting Initiative (GRI) used to measure sustainability in organizations (including industries) and that includes a set of economic, social and environmental indicators (including indicators on labor practices, decent work, diversity in labor, human rights and society/community relationships).

Based on Agenda 21 recommendations, some cities also have a set of sustainability indicators. For example, City of Hamilton, Ontario in Canada reports its sustainability indicators under Vision 2020 program.

The Norwegian Agency for Development Cooperation has specific indicators for assessing economic aspects of institutional sustainability and social sustainability. Ecological sustainability in companies can be achieved through: total quality environmental management, ecologically sustainable competitive strategies, technology transfer and corporate population impact control; and indicators will depend on the goals. A community's ecological priorities can be seen in its choice of indicators. For example, Norwich City in the U.K. has indicators for clean air, domestic waste, water saving, energy saving, clean river water, more wildlife, clean streets etc. Instead of measuring all the indicators (social, environmental and economy OR separate for companies and communities) a few authors have suggested a more integrated or alternative approaches such as 'anthro-capital' that includes human capital, social capital and constructed equity (McElroy, Jorna and van Engelen 2008); systems theory; or sustainability quotient approach (in this case it can be a regeneration quotient approach).

Sustainability quotients express the relation between "… the rate of capital resource consumption and/or production [impact] by an organization with the corresponding rate of capital resource supply, or need, proportionately allocated to the organization in some way [entitlements]" (p. 7, as quoted in McElroy, Jorna and van Engelen, 2008, p. 228). An entitlement in these cases is similar to the concept of 'carrying capacity'. For better planning and to be used in policy-making, indicators need to be well-defined (Reynolds and Cavanagh 2009).

Purpose of Indicators

As discussed by Häkkinen (2007), "In policy making, environmental indicators are used for three reasons: (1) to supply information on environmental problems, in order to enable policy-makers to value their seriousness; (2) to support policy development and priority setting, by identifying key factors that cause pressure on the environment; (3) to monitor the effects of policy responses. Indicators are not only needed in order to supply information about the state and condition and causalities, but also to assess the effectiveness of alternative responses." Some authors (Briassoulis 2001[2]) also see indicators as 'decision support instruments' but there is a problem in the design, comprehensiveness and implied causal relationships of these indicators hence they do not make a significant impact on planning. In some studies the authors have used indicators to study stakeholder relations especially in the field of consultation within affected communities (Nurick and Johnson 1998) but they also see indicators as exogenous to the actual process of policy making (Astleithner et al. 2004, p. 4).

Indicators for Regeneration Projects

While considering indicators for a regeneration project, the indicators should go beyond just measuring construction to include socio-economic indicators, impact on local economy, etc. Usually such projects are seen as construction projects and indicators for a construction company would include project performance and company/organization performance (Park et al. 2008). Some indicators can be added to make sure the buildings being renewed meet sustainability criteria and sustainability quotient assessment methods can be used to come up with indicators for it. Another way is to have a checklist with scores, as shown below in this chapter. However, when urban sustainability is added to the mix, indicators to measure urban sustainability must include an array of economic, social, and environmental aspects, and need to include at least one of the following four characteristics: *integrating* (linking economic, social and environmental problems, e.g., unemployment which is an indicator of an socio/economic problem and indicates social stress), *forward-looking* (a trend by looking at past numbers and establishing a benchmark for future expectations, e.g., keep unemployment rate below 3% for next five years), *distributional, and broad inputs from multiple stakeholders in the community* (this involves surveying local people in the community and the experts in the specific field) (Cubin et al. 2006). Examples for selecting indicators for municipalities as well as builders (LEED list of checklist, some of these indicators can be picked up by builders to adopt sustainability principles into regeneration, see http://www.usgbc.org/leed) are given in the example below.

The growing field of urban regeneration and policy created a need for statistical information that led to various studies such as 'booming towns' (Champion and Green 1990), 'Northern lights' (Breheny et al. 1987), 'quality of life' (Rogerson et al. 1989) and 'measurement of geo-demographics' (Brown and Batey 1994) (Wong 2003, p. 258). All of these studies have raised discussion over different methodological issues to create composite indices.

---

[2] Astleithner et al. 2004, p. 4.

Indicators can be presented in different ways. One way is just a report card or score card on all the indicators with either values or grades! In some cases key indicators within each critical area are combined to come up with a final 'regeneration index' value. It is important to note that in this case each key performance indicator and critical area is weighted differently depending on their level of importance for regeneration in a given area. When the index is being tracked over period of time, weights given to each critical area can be varied if its importance towards regeneration contribution has changed (Cubin et al. 2006).

Calculation of Index: A Simplified Example

To illustrate how an index is calculated, an easy case study is used. The example below shows how the Environmental Quality Index was calculated in the 60s and 70s. There were seven different components/resources (soil, air, water, living space, minerals, wildlife, and timber) that were taken into consideration to calculate the environmental quality index. Table 7.1 shows their score and relative importance (relative importance, also known as weight, is selected by a group of experts including environmental scientists, policy makers and statisticians):

Table 7.1. Scores and Relative Importance.

Category	1970 Score	Relative Importance
Soil	77	31
Air	32	20
Water	42	20
Living Space	58	12
Minerals	48	7
Wildlife	51	5
Timber	76	5

To calculate Environmental Quality Index (EQ), the score is multiplied by the relative importance and the total is divided by 100 to get the EQ points (Table 7.2). The higher the EQ index the better is the state of environment.

Table 7.2. EQ Points.

Category	1970 Score	Relative Importance	EQ Points
Soil	77	31	23.87
Air	32	20	6.40
Water	42	20	8.40
Living Space	58	12	6.96
Minerals	48	7	3.36
Wildlife	51	5	2.55
Timber	76	5	3.80
National EQ Index			**55.34**

Although, more complicated systems to calculate index have been developed, this is just one simple example to illustrate how to carry out the calculations.

What it means is that one looks at one composite number to know the environmental quality instead of looking at seven different numbers and carrying out certain calculations.

Monitoring Systems

There are many indices to measure performance of a construction project. For example, one by the Construction-Industry Institute (see Fig. 7.2 below), includes six categories: cost, schedule, safety, changes, rework and productivity. Others, such as Kaplan and Norton recommend use of a Balanced Scorecard (see Fig. 7.3 below), which has four main categories: financial perspective, customer perspective, internal business processes and organizational learning. The narrative below gives some examples of these monitoring systems and how other places have selected such indicators (Park et al. 2008).

One of the ways of measuring and tracking regeneration is Barry's Point System, which involves collecting data for many regeneration indicators such as transportation, community benefits, etc. In the point system, critical areas are identified that have the greatest impact on the community and indicators to track regeneration goals are aligned with these critical areas. For example, if the goal is to move from private to public transportation in an urban setting, the critical area would be 'transportation and mobility' and indicators could

Figure 7.2. Construction Industry Institute.

Figure 7.3. Kaplan and Norton.

be: how many people own a car, how many people travel by bus or rail, and how many people walk! By comparing baseline information over years, a trend can be established to see how indicators worked towards achieving the goal (or target, if one is established) (Cubin et al. 2006). The way the weighting system works for an area can be seen in the following example (Table 7.3) (Cubin et al. 2006):

Table 7.3. Critical Areas and the Weightings Applied to Each Component (Barry, 2004 quoted in Cubin et al. 2006, p. 14).

Components	Weighting (%)	Ranking
Economy and Work	21.5	2
Resource Use	17.5	5
Buildings and Land use	18.9	4
Transport and mobility	22.1	1
Community benefits	20.0	3

Barry's Point System has many benefits because it integrates information from multiple indicators, to take into account the various factors that affect regeneration and also gives weighted importance to the critical areas. However, the point system does not always work for the urban renewal projects because urban regeneration that comes with growth to an area might look at different types of growth. The point system only looks at the five critical areas identified in Table 7.3, however, if the focus of the regeneration project is to reduce the crime rate in an area, it is not looking at economic growth in which case the critical areas and indicators will be different from the ones identified in the Point System (Cubin et al. 2006).

Examples of some indicators of some cities in Europe and ways to measure and present them (Colantonio and Dixon 2009):

The 10 dimensions, shown in Table 7.4, provide the assessment areas linked to a scoring system—points are awarded from 1 to 5 depending on inclusion of specific items in the regeneration project. "The scoring system can be applied either *ex-ante* to evaluate project proposals through the application of checklists or *ex-post* through the evaluation of performance indicators, which have been selected to monitor the overall progress of the project for each areas." A score of 5 would mean that the project would address issues comprehensively. At the monitoring stage, a score of 5 would mean that the project is performing well and reaching or going above pre-fixed targets (Colantonio and Dixon 2009). The scores can be represented in a radar diagram (Colantonio and Dixon 2009).

Table 7.4. Main Social Sustainability objectives of regeneration schemes (p. 75).

City Theme	San Adrià de Besos/ Barcelona	Cardiff	Rotterdam	Turin	Leipzig
Housing and environmental health	✓	✓	✓	✓	✓
Education and skills	✓	✓	✓	✓	✓
Employment	✓	✓	✓	✓	✓
Health and Safety	✓	✓	✓	✓	✓
Demographic change (aging, migration and mobility)	✓	✓	✓	✓	✓
Social mixing and cohesion	✓	✓	✓	✓	✓
Identity, sense of place and culture	✓	✓	✓	✓	✓
Empowerment, participation and access	✓	✓	✓	✓	✓
Social capital	✓	✓	✓	✓	✓
Well being, Happiness and Quality of Life	✓	✓	✓	✓	✓
Equity	X	X	X	X	X
Human rights and gender	✓	X	X	✓	✓
Poverty	✓	X	✓	X	X
Social justice	X	X	X	X	✓

The radar diagram (Fig. 7.4) shows that the project scores very high in meeting softer sustainability themes such as empowerment, participation and access but other areas such as employment, education and skills, etc. can be improved.

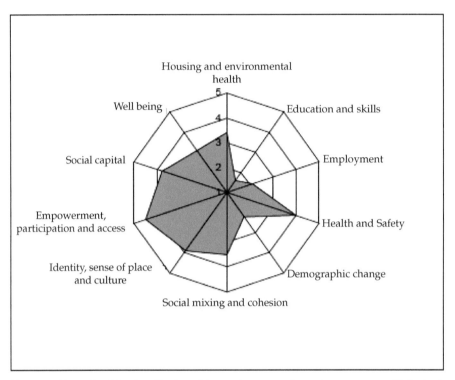

Figure 7.4. Example of the visualization of the scoring system for project assessment (Colantonio and Dixon 2009, p. 76).

Another way of looking at these indicators is by aggregating them into a 'composite index' which is similar to a 'social index' which gives an idea about the social sustainability performance of either a project or a place. In this case, weight and aggregation methodology used will depend on the priority of various social objectives in the regeneration scheme and also in the context where the assessment is being carried out. Then the overall score can also be categorized into different performance levels—e.g., gold, silver or bronze or a 'traffic light' assessment system. In this case green light represents an area that is performing satisfactorily, yellow light indicating an area that needs some attention while red color signifies an area that has overall unsatisfactory results and needs attention. The results shown in the radar diagram (Fig. 7.4) can be depicted in Table 7.5 using categories of performance and the traffic light system (Colantonio and Dixon 2009).

The assessment framework and scoring system described above helps to assess social sustainability measures for regeneration projects and has both advantages and disadvantages. Some advantages include: the framework provides guidelines on the main themes and principles of sustainable urban

Table 7.5. Scoring System.

Theme/Dimension	SSAF score	Result	Traffic light system	Aggregated scores
Housing and Environmental health	3.4	Good		3.2
Education and skills	1.6	Very poor		
Employment	2.1	Poor		
Health and safety	4	Good		
Demographic change	2.2	Poor		
Social mixing and cohesion	3.6	Good		Barely
Identity, sense of place and culture	3.9	Good		Acceptable
Empowerment, participation and access	4.5	Very good		
Social capital	3.8	Good		
Well being	3.1	Barely acceptable		

regeneration which can be implemented by policy makers or practitioners; sub-themes are flexible and can be adopted to the context it is being used in; and the framework can be based on a normative model. Some of the limitations of the framework include: it is over simplified; too many indicators may limit the full operationalization of the framework; and the assessment has limited stakeholder participation in the selection of indicators and checklists (Colantonio and Dixon 2009).

How to Apply Monitoring and Evaluation Framework

Monitoring and evaluation is an important step to highlight challenges and to improve the effectiveness of either an organization/builder or government so that regeneration impact can be achieved. The monitoring and evaluation team should aim to deliver: performance information and analysis both at a program level (applicable for governments) and project level (applicable for both governments and builders); monitoring and evaluation of outcomes and impact across the whole government; and projects that seek to improve monitoring and evaluation performance (mainly in government monitoring). Overall monitoring and evaluation should contribute to improved governance; efficient and effective projects with an impact; should be development oriented and methodologically sound (Umtha Strategy Planning and Development Consultancy).

For the purpose of reporting, a framework is needed—generally, a results based framework is recommended which clearly shows linkages between the objectives to be achieved (by an organization or ministry) and the results. The following points need to be addressed to have a results based framework for monitoring and evaluation (Umtha Strategy Planning and Development Consultancy):

- Strategic Objectives: Strategic objectives are usually higher level goals for an organization.
- Results: In a results based framework for evaluation, at least one (but could be more than one) target result is set for a strategic objective. Results are, in a way discrete, outcomes that are expected in a project or series of projects. For the monitoring and evaluation process then it is critical that assumptions made for achieving the results are also stated, since any change in the assumptions can influence the results.
- Indicators: Indicators for both strategic objectives and results need to be identified. In an organization these would be directly linked to the activities of the organization. Just like indicators are needed for the project, Key Performance Indicators (KPIs) should be set for each person responsible for certain actions to measure small steps that will contribute to larger objectives.
- Means of Verification: To show success for a set of indicators, data over a period of time is needed. As part of the process for monitoring, defining what data will be needed (so that it can be measured) should be done upfront. This will help data collection as the project goes along, instead of trying to find information at the last moment.
- Project Planning Matrix: For each project, a one page summary matrix with all the topics discussed so far (objectives, project purpose, results, activities, assumptions and indicators) becomes an important planning and reporting tool.
- Dashboard and Monitoring Framework: This is a reporting template and some examples are shown below with indicators and how they can be reported.

As a first step to select indicators, it is important to define the vision for the renewal project (that should synchronize with the vision of the city) (Cubin et al. 2006, p. 24).

Before setting a vision and strategic objectives, public engagement to carry out neighborhoods' analysis (and introduce sustainable development concepts in regeneration) is important. The process that can be followed is shown by the 'Sustainable Renovation of Buildings for Sustainable Neighborhood'.[3]

---

[3] http://www.eukn.org/France/fr_en/E_library/Housing/Housing_Quality/Sustainable_Construction_Methods/HQ2ER_Sustainable_renovation_of_buildings_for_sustainable_neighborhoods.

The working group holds a great significance in neighborhood regeneration projects. In each neighborhood, the working group is accompanying municipalities through the main phases of their neighborhood regeneration projects, i.e., neighborhood analysis (inventory, diagnosis, and setting of strategic priorities), action plan working out (generation and choice of scenarios) and implementation and monitoring (projects upon the open spaces and on buildings and their monitoring. The working group is helping municipalities introducing sustainable development in each regeneration projects phases:

- By ensuring a neighborhood shared diagnosis for Sustainable Development (SD), a setting of SD priorities and SD objectives.
- By including SD in the generation of scenarios and in their assessment.
- By ensuring the working out and the implementation and monitoring of SD projects.

The working group then tackles the following problems in a sustainable manner:

- How to foster residents and users participation.
- How to preserve and enhance built and natural heritage and conserve resources (energy, water, land space, materials).
- How to improve the quality of the local environment (housing and building quality, risk management, air quality, etc.).
- How to ensure diversity.
- How to improve integration of the neighborhood in the city.
- How to reinforce social life.

After setting a vision, the next step is to identify the critical areas that are important to regeneration efforts in an area/city/region (e.g., 'economy and work', 'transportation and mobility', 'housing', etc.) (Cubin et al. 2006, p. 24). As discussed by Nel-lo (2010), the first lesson learnt from the various urban renewal projects is that there is a "need for comprehensive vision" for more equitable growth.

"Critical areas of regeneration are broad components from the economy to the environment which are perceived to be the areas needing improvement. These are also specific areas which are important to monitor, because their positive or negative impacts will help to gauge the overall regeneration" (Cubin et al. 2006, p. 24). As defined by Colantonio and Dixon (2009), urban regeneration projects can generate potential outputs and outcomes in at least 10 social sustainability dimensions and policy areas (which are also critical areas for the social sustainability of local communities and neighborhoods): demographic change; education and skills; employment; health and safety; housing and environmental health; identify, sense of place and culture; participation, empowerment and access; social capital; social mixing and cohesion; and well-being, happiness and quality of life (Colantonio and Dixon 2009, p. 4).

Master plans, development plans and public meetings can be used to determine critical areas. The next step is to choose key performance indicators for each of the critical area(s) identified (e.g., for 'economy and work' key indicators can be unemployment or average income). Data for indicators can be taken from census, annual monitoring reports, national statistics, and local surveys (Cubin et al. 2006). Thus, the framework of deriving key performance indicators consists of three steps (Park et al. 2008):

STEP 1: Translate vision into strategic objectives

For any urban regeneration project, defining the key objective(s) is important—in most cases the main objective is to recreate old urban areas and rehabilitate urban districts with regional socio-cultural characteristics (Park et al. 2008). Regenerators should focus on how much a project (based on (Park et al. 2008)):

- is consistent with national development plan, level of economic efficiency and degree of sustainability;
- is inclusive of public demands, diverse functional spaces and cultural spaces with green land; and
- contributes to revitalization and promotion of attraction.

In the examples section, an example from Cape Town reflects how urban renewal program's strategic objectives were aligned with City's strategic plan.

STEP 2: Define critical success factors for every strategic objective

The critical success factors for an urban renewal project can be grouped into five categories: customer, finances, execution/implementation process, innovation and sustainability (Park et al. 2008).

STEP 3: Define key performance indicators for each objective/critical success factor: Performance indicators are then based on each critical success factor. An example of how inclusive the project is to community needs can be measured by how many public meetings were held! Some examples of indicators and how they can be represented are given in the annexes below. One of the examples is HQE2R 'Integrated Sustainable Development Indicator System' (ISDIS) which gives a set of core indicators based on experience in a few European countries. ISDIS gives five sustainable development objectives, 21 targets system with their 51 key issues and 61 indisputable indicators.

Example of how to match goals with indicators and refine the indicators of success (from East-Thuringia, Germany)

East-Thuringia was undergoing sustainable spatial development at a regional scale and to measure the progress two steps were taken (Birkmann 2003, p. 300):

1. relevant goals and targets of sustainable spatial planning were identified and this was used as a basis to select indicators.
2. the available data was used to identify measurable indicators.
   These goals and settlements are shown in Table 7.6.

Table 7.6. Indicators used for measuring sustainable spatial development at the regional level.

Goals for Settlement	Indicator
Reduction of resource-intensive settlement and traffic structures	Development of settlement and traffic space (development in %)
Containing urban sprawl	Development of settlement and traffic space in high order centres in relation to medium and lower order centers and the rural area (linked to the classification of the central place system)
Promoting cycles of land-use	(a) Amount of brownfields in the region (total amount in ha) (b) Percentage of reused area of brownfields (percentage of the total brownfield area)
Suitable supply of employment and jobs	Unemployment rate (percentage of the unemployed of all employees)
Contentment in regard to regional conditions	Internal (regional) migration rate (balance of inland and regional migration per 1000 inhabitants)
Goals for Open Space	
Protection of endangered species	Development of the population of key-species (indicating the situation of specific habitats)
Protection of open space	Degree of fragmentation (different methods, for example the percentage and size of open space which is not fragmented by national and state roads/railways)
Goals for Networks	
Ensure mobility for all	Kilometre performance of public transport (vehicle-km per subregion per year)

Source: Birkmann 2003, p. 301.

As discussed by Birkmann (2003, p. 301), development of brownfields became an integral part of regional planning. However, the definition of brownfields and tools to measure progress towards development of brownfields were inadequate for a long time. Finally two main indicators for reuse of industrial and commercial brownfields were defined. The first indicator requires recording the number of existing industrial and commercial brownfields that exist, which gives an indication of the relevance of the issue for the region. The second indicator then focuses on the measurement of the percentage of developed brownfields (which means how many former brownfields are being currently used and for what purpose). Since the indicators focus on processes, it is necessary to measure them over a period of time to indicate success. The second indicator is then divided into three levels based on reuse rate—low (0–40% reuse), middle (41–79%) or high (80–100% reuse). Initially the indicators were mainly used as information tools or to observe change in economic or environmental conditions, they were not really used in planning or to influence policy and planning. The inclusion of indicators in policy and planning have led to changes in the types of indicators selected to analyze the efficiency of different planning instruments, this means the purpose of the indicator has changed from a descriptive to a more pro-active approach. Since planners are always confronted with how to balance the deep-rooted social, economic and environmental conflict and which indicators to use, it is important to link the indicators with planning goals as well as to focus on the target group that is going to use the tool. There are certain limitations on measuring sustainable spatial development or urban regeneration, but the focus should be on improving monitoring, controlling and evaluation instruments.

Another example to show Alignment of Strategic Objectives of Renewal Program with City's Strategic Plan

It is important that the urban renewal program's strategic objectives should align with City's strategic plan's objective. One such example is from Cape Town, South Africa, where they have aligned council's strategic focus area with urban renewal program strategic objectives as shown in Table 7.7 (Umtha Strategy Planning and Development Consultancy):

Table 7.7. Strategic focus, objectives and indicators in Cape Town, South Africa.

Council's Strategic Focus Areas	Urban Renewal Program Strategic Objectives	Indicators
Shared Economic Growth and Development	Promoting Local Economic Development to reduce poverty and unemployment	• % increase in area contribution to GDP growth • % increase in household income levels • Number of jobs created • % increase in investment in the area • % decrease in the rate of unemployment
Public Transport Systems	Creating job opportunities through the Extended Public Works Programme (EPWP) Developing efficient, integrated and user friendly transport systems	• % increase in employment levels • % increase in skilled people • % increase in capital projects conforming to EPWP principles • Km of road built • Public transport promotion • Development of residential areas next to public transport opportunities
Safety and Security	Providing a safe and secure environment by fighting crime	• % decrease in crime levels • % decrease in gangsterism • % decrease levels of substance abuse • Increase in recreational opportunities for the youth

Table 7.7. contd....

Table 7.7. contd.

Council's Strategic Focus Areas	Urban Renewal Program Strategic Objectives	Indicators
Health, Social and Human Capital Development	Delivering well managed safety nets (social cohesion)	• % increase access to grants • % increase to indigence support • % increase to services (water, electricity, sewage, telecoms, refuse removal)
	Supporting education, training and skills development	• % increase in number of people with matric and post-matric qualifications • % increase in number of people skilled • Increase in number of training opportunities • % increase in culture of learning and teaching
Sustainable Urban Infrastructure and Services	Creating a quality urban environment where people can live with dignity and pride	• % increase in place making interventions • % increase in development or enhancing of public open spaces

In the end for monitoring and evaluation to be part of the project or program, a champion is a must; who believes in it to improve the effectiveness of the organization; a certain amount of budget needs to be allocated to the process; and a communication plan is needed to convey the success of the project with a complete set of indicators to support it.

Conclusion

"What gets measured gets done" has been quoted so often that it has been attributed to different authors including Peter Drucker, Tom Peters, Edwards Deming, Lord Kelvin and many others.

Consequently, we get hung up on metrics and measuring things to the point that we sometimes lose track of measuring what really matters?[4]

It is important that a vision and strategy for a regeneration project be established, followed by the selection of the right set of indicators and how to monitor and analyze them! It is also important that the indicator-based monitoring and controlling instruments are based on scientific criteria (including validity of data, transparency of selection criteria and methodology used). As mentioned by (Birkmann 2003, p. 299). In this context, measuring tools and indicators should:

• focus on a target group;
• be linked to goals and targets of spatial planning;

---

[4] http://www.swspitcrew.com/articles/What%20Gets%20Measured%201106.pd.

- identify the functions they serve;
- be based on data which are readily available;
- only measure important key-elements instead of trying to indicate all aspects;
- integrate macro- and project-oriented indicators;
- look at trends and developments.

It can also be argued that it is not always easy to define the cause-effect relationship in the complex interdependent world. For example, as discussed by Wong (2003), it can be argued that the driving force of urban change/ regeneration can be due to structural changes and historic inertia at the local level, or because of some external factors from national and global forces, or it can also be an interaction of internal and external forces. This makes it almost impossible to separate the dynamic process of change from the state of outcomes, because these states of outcomes can also lead to further changes. This means that urban change or regeneration encapsulate the process of change as well as the state of performance, which makes it difficult to really make a causal input-outcome model of measurement difficult (Wong 2003, p. 261). As further discussed by Wong (2003, p. 262–263), some other problems in connecting theory and measurement in practice include: the relationship between input factors and outcome phenomena is not that easily identifiable because it is non-linear and ambiguous; indicator values cannot be always interpreted because some factors have ambiguous relationships with the social phenomenon that is being measured; and it is difficult to define the appropriate timescale to prove the causal relationships. Although some of these concerns do exist in making the decision to choose the right set of indicators but the choice of indicators to measure the right things and influence policy cannot be over-emphasized. A badly designed set of indicators can cause a tremendous damage to show progress of ongoing project, public debate or a policy debate (Wong 2003, p. 267).

References

Bankes, N. and A.R. Thompson. 1980. Monitoring for impact assessment and management. Westwater Research Centre, Vancouver.

Birkmann, J. 2003. Measuring sustainable spatial planning in Germany: Indicator-based monitoring at the regional level. Built Environment 29(4): 296–305.

Carley, Michael J. 1986. From assessment to monitoring: Making our activities relevant to the policy process. Impact Assessment 4(3-4): 286–303.

Charot-Valdier, Catherine, Philippe Outrequin and La Calade Celia Robbins. 2004. The HQE2R toolkit for sustainable neighborhood regeneration and European application overview. March 2004.

Colantonio, Andrea and Tim Dixon. 2009. Measuring Socially Sustainable Urban Regeneration in Europe. Oxford Institute for Sustainable Development.

Cubin, Nicholas, Jeffery Doyon, Nathan Malatesta and Christopher Warms. 2006. Wembley Regeneration Index: The Application of a Point System to Measure the Continuous Regeneration in London, Borough of Brent.

Grover, Velma I. 2001. Environmental Indicators: An overview. ISWA Times, Denmark, Issue 3.

Häkkinen, T. 2007. Assessment of indicators for sustainable urban construction. Civil Engineering and Environmental Systems 24(4): 247–259. http://new.usgbc.org/resources/new-construction-v2009-checklist-xls.

Isabel Breda-Vázquez, Paulo Conceição and Pedro Móia. 2010. Learning from urban policy diversity and complexity: Evaluation and knowledge sharing in urban policy. Planning Theory & Practice 11(2): 209–239.

Nel-lo, Oriol. The challenges of urban renewal: Ten lessons from the Catalan experience. Análise Social, Vol. 45, No. 197, Urban Governance in Southern Europe (2010), pp. 685–715. Published by: InstitutoCiênciasSociais da Universidad de Lisboa.

Park, Heedae, Du Yon Kim, SeungHeon Han and Sang H. Park. Approaches for Performance Measurement of Urban Renewal Mega Projects. The 25th International Symposium on Automation and Robotics in Construction, Institute of Internet and Intelligent Technologies, June 2008. Vilnius Gediminas Technical University, Sauletekio al 11, 10223 Vilnius, Lithuania, http://www.isarc2008.vgtu.lu.

Pinter, Laszlo. Measuring Progress Towards Sustainable Development Goals. International Institute for Sustainable Development. January 2013. http://www.iisd.org/pdf/2013/measuring_progress_sus_dev_goals.pdf.

Reynolds, P. and Rob. Cavanagh. 2009. Sustainable education: Principles and Practices. Annual Conference of the Australian Association for Research in Education, Canberra.

South African report Umtha Strategy Planning and Development Consultancy, Monitoring and Evaluation Framework: The Urban Renewal Program.

Wong, Cecillia. 2003. Indicators at the crossroads: Ideas, methods and applications. The Town Planning Review 74(3): 253–279.

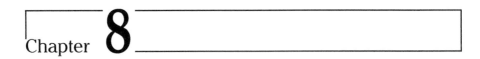

Chapter **8**

Valuation and Job Outlook after Regeneration

Introduction

Many of the common inherent problems in urban decay (as discussed in earlier chapters in this book, and also captured by Roberts's (2000, p. 16) and Otsuka and Reeve (2007)) require policy responses that can generally be classified under six main thematic areas:

- the relationship between poor physical infrastructure conditions and social deprivation;
- the continuous need for physical replacement of old urban fabrics;
- the strong links between economic success and urban prosperity;
- the drive to maximize beneficial uses of land while avoiding unnecessary urban sprawl;
- the changing political climate and priorities expressed in changing urban policies; and
- the emerging and growing importance of the sustainability agenda.

The six thematic areas above are anchored on dealing with the urban decay problems within a three pillar approach with a special focus on buildings (physical), unemployment, social problems (economic and social) and degrading environmental issues. Within this perspective urban regeneration planning is conceived as part of a sustainability approach (economic, social and environmental) as defined by Roberts (2000, p. 17): More specifically it calls for ..."comprehensive and integrated vision and action which leads to the resolution of urban problems and which seeks to bring about a lasting improvement in the economic, physical, social and environmental condition of an area that has been subject to change."

To be effective and to create sustainable and inclusive communities after the renewal process, it is important to have complementary and coherent economic, social and environmental regeneration initiatives. Thus, regeneration should cover a broad range of public policy. The 3Rs Guidance (ODPM 2004) defines regeneration as being *"a holistic process of reversing economic, social and physical decay in areas where it has reached a stage when market forces alone will not suffice"*. The British government in 2010 also looked at the holistic definition of regeneration and now sees it as a way to redistribute wealth *"regeneration can help us make the best of our assets and our people. It can help areas adapt to new roles, and improve the distribution of wealth and opportunity. It can restore social justice, and reduce community tensions. And as the country adapts to a smaller state, regeneration can play a vital role for communities, by fostering a sense of solidarity and hope"* (Ministerial Statement at the National Regeneration Summit, 14 July, 2010).

The three pillars of regeneration typically focus on the four key aspects of regeneration, as shown in the Fig. 8.1.

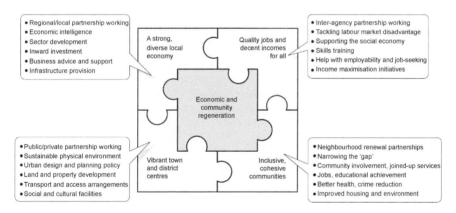

Figure 8.1. Four Key Aspects of Regeneration.
Source: Audit Commission, nd.

This chapter looks at how to value (valuation) regeneration and the types of jobs available during and after regeneration activities. It describes linkages between the level or type of regeneration activities carried out—i.e., was it community level, commercial or industrial type regeneration and investment made in training the local unemployed youth for available jobs.

While planning regeneration activities, it is also important that emphasis is on diversification of economy and jobs available (this has become more important in the era when prices of commodities fluctuate widely impacting the local economy, e.g., oil, timber, uranium). Essentially this means that more than one type of income source should be planned for the city or region undergoing regeneration.

Benefits of Regeneration

Essentially regeneration is about closing gaps. When a three pillar approach (economic, social and environment) is used in regeneration, the process of economic regeneration is concerned with delivering impacts on targeted regeneration areas or particular groups in society (e.g., unemployed local people) to enhance their prospects. In most of the places where equity and diversification are to be achieved, governments need to intervene. As discussed in the HM Treasury (2008) (p. 51), the rationale for intervention on the part of Government has been heavily influenced by the need to overcome market failure and the achievement of an equity objective or any other broad social objectives, such as local or regional regeneration. In this case (especially when government investment is involved), it has been argued that the successful regeneration is "about achieving additional economic, social and environmental outcomes that would not otherwise have occurred (or which would have been delivered later or of a lower quality) whilst also representing good Value for Money for the public investment" (Tyler et al. 2010).

Generally the results of regeneration are both **direct** (which includes benefits for people or places where regeneration has occurred) as well as **indirect** effects (indirect effects benefit society as a whole and include things such as better work opportunities, income opportunities, improved health and reduced crime rate). It is important that these indirect benefits are valued though it is not always easy to quantify these indirect benefits and show the extent to which these indirect effects arise from regeneration activities.

Defining the Logic Chain from Inputs to Outcomes

There are different ways of valuating regeneration and in doing so valuation methods need to be considered. Essentially, this means that for all the regeneration activities, it ideally needs to specify a logical pathway from **inputs through activities to outputs, outcomes, impacts and value.** These pathways then reflect the 'theory of change', which depicts how different types of regeneration activities are bringing about change for people or places in local regeneration areas. The links between the individual elements are usually spelt out in a 'logic chain' (Tyler et al. 2010).

Valuation Techniques and Job Skills

A central element of all approaches to valuing the costs and benefits of regeneration policy is to identify 'who will benefit' from these regeneration activities although it is not always easy to identify the beneficiaries. This is because, at times, regeneration just has direct benefits and beneficiaries, so for example if a building or a place undergoes regeneration activities, then it could be the people who live in that building or place (have direct and strong correlation between the intervention and who benefits). But at times there

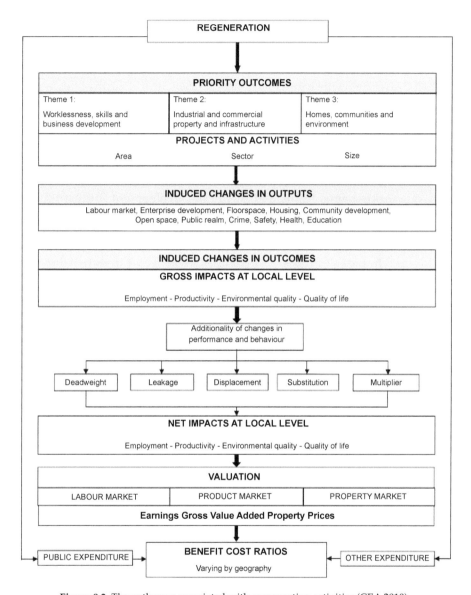

Figure 8.2. The pathways associated with regeneration activities (CEA 2010).

are also people who indirectly benefit from it—for example, people who visit the place and benefit from the improved quality of the place or even if they are just passing through this place (in this case there is a weaker correlation between intervention and beneficiaries).

Generally, regeneration activities will have impact on a diverse range of individuals across society with considerable variation by income, gender, ethnicity, age, geography and disability. Because this variation impact of regeneration will also vary on these ranges of individuals. This means that efforts should be made to identify and quantify the distributional effects of these regeneration policy interventions. One of the ways to do so is to establish the value associated with a **unit** of regeneration benefit, the **worth** of this benefit may be greater to those with lower incomes who tend to be disproportionately concentrated in the most deprived areas. Making these adjustments (to account for distributional effects) in the Benefit cost ratios give a more accurate picture of the benefits (Tyler et al. 2010).

Usually when looking at the economic valuation of regeneration activity, categories and types are identified under **three main themes** of regeneration activity: **Worklessness, Skills and Business Development; Industrial and Commercial Property and Infrastructure; and Homes, Communities and the Environment.** Unit cost for a given level of public sector expenditure is estimated and then the volume of net additional outputs generated is calculated. The next step is then assigning monetary value to each of these net additional outputs followed by a cost-benefit analysis (Tyler et al. 2010, p. 13).

With economic regeneration, the types of job created will depend on two parallel activities that need to take place (and also where the regeneration activities have taken place):

1. Worklessness, skills and training: This step would include providing information and guidance to unemployed people about the employment opportunities available after regeneration, help with job search activities and those activities that help gain employment such as subsidized education and training programs to enhance skills to improve employability. Some other actions can include creation of funds to aid these activities. Examples here include Working Neighborhoods Fund or Neighborhood Renewal Funding in U.K. The first step should focus on reaching out to people with information about what is available in the market, what are the options open to them, providing them with advice and guidance on how to become employable and giving them confidence to join the work force. The second step will then involve retraining people to give them skills to be employed again, helping people then find jobs using their new skills and helping them stay at work (one of the ways to do so could be by giving incentives to businesses to not just hire for a short term or for a supported internship but employ for a longer term). The intention is to increase the employment rate and move people away from social benefits (unemployment or social insurance). "While these benefits need to be treated separately from real resource benefits to the economy from incomes and Gross Value Added, they are a legitimate consideration for those appraising and evaluating the performance of regeneration interventions" (Tyler et al. 2010, p. 14). There are some issues in using this approach and the factors stated above for valuation because it involves issues of labour demand, and it is

difficult to attribute what skills and training were provided by whom to whom (which funding grant helped retraining a person) and the impact of interventions (if the person had earlier training as well, what is the value and role of retraining) and other developments and programs mounted independent of the intervention by government through the regeneration program. Another issue is that of the timescale—training people (or a group of people) and getting them sustainable employment is not a short term activity (it could be a matter of months(s) if not years), what is then the reasonable time period over which benefits of particular interventions and economic regeneration activities can be claimed? Some of the approaches used to evaluate how regeneration activities enhanced individual skills and earnings include: interviewing beneficiaries, and/or assessment of additional jobs by using cost benefit analysis. Some studies in Scotland have also tried to provide evidence on the unit costs of intervention (i.e., costs per beneficiary) and evidence on the net addition associated with the delivery of key outputs (e.g., positive outcomes into employment, or improvement of skills) (Tyler 2010, p. 19).

Economic Regeneration aims "to change behavior, encouraging and supporting local people to become more entrepreneurial. Possible priorities for economic regeneration are: employment for all, improving employment rates, helping the hardest to reach into work and building a stronger local economy." This can be done via specific programs such as helping start up businesses, skills training, and encouraging tourism and creativity projects. However, as studies by Professor Campbell (Leeds Metropolitan University) show that it's not always necessary that economic regeneration brings jobs to unemployed local people, it is important to have an employment strategy that would benefit the local people. "A review of actions to help tackle worklessness shows they include actions to tackle both the 'supply side' and the 'demand side', as follows:[1]

Supply Side:
- Increasing access to employment by improving transport links and providing accessible childcare
- Increasing skills base by providing education and training
- Helping the 'hardest to reach' into work
- Improve support services to lone parents
- Increase support to 'second generation' workless
- Deliver support services differently, e.g., via libraries, employment support centers, mobile job support centers
- Providing intensive outreach and support to help overcome a culture of worklessness.

---

[1] Key Findings of the Economic Scrutiny Committee Short Scrutiny Study—Economic Regeneration; Welsh Case Study: 1994–2003.

Demand Side:

- Challenging employers' behavior, e.g., why aren't they interviewing unemployed applicants
- Encouraging a suitable infrastructure to support new job creation, e.g., transport, housing choice and access to good quality services
- Providing accommodation for new business
- Attracting and creating new jobs
- Encourage business start-ups
- Providing business advice and support
- Forming a partnership with local employers
- Providing access to broadband."

2. Enterprise and business development: Most of the regeneration initiatives under this theme focus on improving the economic wellbeing of a neighborhood and involve actions such as general support for business growth (including competitiveness), assisting start-up businesses, trade promotion and also promotion of enterprise, research and development (Tyler 2010, p. 21). The idea is that by supporting and promoting local businesses they will be able to employ local people and help improve the economic and social wellbeing in the area.

3. Regeneration activities in housing usually would mean construction of new houses and demolition of old ones. It is rare that improvements are done to the existing houses. This does give an opportunity to develop modern, eco-homes (improve the quality of living area), or change the residential mix. Although, homelessness is not a main focus of regeneration but if homelessness is a problem then the focus can also be on more affordable housing or supporting vulnerable groups.

4. Regeneration activities happen in declining neighborhoods (as discussed above) it can also occur on industrial or commercial properties. The second category includes activities such as land reclamation, developing brownfields and greenfields, site servicing and maybe facilitating the development of new industrial or commercial floor space. These activities are heavily supported by government funds usually in partnership with the private sector. Valuation in this case can be done either by looking at the production benefits associated with the end use of industrial or commercial property linked to employment or gross value added or by valuation of the property asset itself (Tyler 2010, p. 30 and 31). Different approaches that can be used for valuation include hedonic price models, Delphi technique, cost benefit analysis or tracking of property investment returns and vacancy chain analysis. Some studies have also looked at total returns index (Adiar et al. 2003, Adiar et al. 2005) or even residual 'stigma' associated with the remediation of contaminated land (Bond 2001) (p. 34). One of the major challenges in valuation is availability of data "...

the appraisal of urban regeneration sites is one of the most challenging tasks currently facing the valuation profession" (Tyler 2010, p. 35).

An Example to Show Inclusion of Training and Employment as Part of Regeneration Activities in UK

Initial urban regeneration policy in the UK (1980s) focused on property development based on a trickle down approach. This assumed that if offices and factories are built in an area with higher unemployment then the local unemployed residents would get an opportunity to find employment. However, it was soon realized that this theory is not working (Centre for Local Economic Strategies 1992). Just building factories in the areas of higher unemployment does not guarantee jobs for the locally unemployed because they may not have the right skills. This started the era of targeted training initiatives that were aimed to train local residents to become more employable or compete for the available jobs. Since the construction projects were the most tangible sign of regeneration activity, focus also shifted on making the contracts[2] in a way that the contractors have to hire certain number of local residents. However, the use of untrained local people was not always effective. All the attempts of improving this effectiveness were shortened by the introduction of the Local Government Act of 1988. The Act however, does provide guidance for alternative ways of recruiting local residents (Department of the Environment 1993).

During the 1990s two City Challenge initiatives were announced (11 challenges ran from 1992 to 1997 and second round from 1993 to 1998) under the Single Regeneration Challenge Fund. The Challenge, aimed at regeneration of inner cities, was a partnership between local authorities, local communities and the private sector. Most of these areas had high levels of unemployment, poor environmental and housing conditions and other indicators of multiple deprivations. Considerable reliance of these regeneration initiatives were put on physical development but the idea was to have urban regeneration management approaches that would link physical infrastructure, residents and the wider economy. Studies have also shown that a number of local residents were employed in construction and associated activities, and studies have also shown that of the 20 Challenges (who participated in a survey), about 80% ran employment recruitment services as well as training initiatives targeted at the construction industry for the local residents. Table 8.1 shows that maximum expenditures were on construction and only a small percentage was spent on training.

---

[2] Contracts were of various types: compulsory agreements (where contractors were committed to recruit certain proportion of labour force from local residents), negotiated agreements (rather than competitive bidding of tenders, contract was awarded to selected contractors that agreed to hire local residents and public agency invested in training such local residents to do the job), and voluntary agreements (where contractors agreed to interview the local residents and hire them if they are suitable for the jobs. So there was no commitment to hire local residents).

Table 8.1. Percentage of spending over five years allocated to specific categories in selected City Challenges.

City Challenge	Percentage of five-year spend allocated to specific headings (row percentages)				
	Physical works[1]	Training/Business development	Housing[2]	Community[3]	Management[4]
Bradford	20.0	16.5	41.6	18.2	3.7
Deptford	22.9	14.8	51.2	7.5	3.6
Hulme	19.6	8.0	58.3	10.5	3.6
Kirklees	28.1	29.1	29.7	9.1	4.0
Liverpool	50.4	24.9	16.7	4.2	3.8
Middlesborough	18.8	13.3	32.5	31.7	3.7
Newcastle	15.7	10.3	38.9	30.2	4.9
Nottingham	42.0	21.1	7.6	26.6	2.7
Stratford	32.0	16.0	32.0	14.7	5.3
Walsall	45.7	15.5	15.5	19.6	3.7
Wigan	34.0	18.2	27.0	17.3	3.5
Wirral	44.5	9.4	16.3	26.0	3.8
Wolverhampton	33.4	17.1	25.4	20.4	3.7
AVERAGE	31.4	16.2	30.2	18.3	3.9

Notes:

[1] Physical expenditure includes expenditure upon commercial and industrial development (new build and refurbishment, road construction and associated infrastructure and land reclamation).

[2] Housing expenditure includes new build and refurbishment, associated environmental works and in some cases expenditure upon management initiatives.

[3] Community spending includes a range of projects such as security, child care, support to voluntary groups, and the provision of sports facilities. The latter accounts for considerable expenditure in some Challenges.

[4] Management covers staffing and associated costs.

Source: Derived form the Challenges' five-year Action Plans.
Source: Hayton 1998.

The main objective of the Challenge initiatives was to promote employment and training opportunities for the local residents of the areas where regeneration activities were carried out. Studies have shown that the geographical focus area was very small and it did not have enough people execute a credible recruitment initiative. A secondary objective was to get construction work for local building companies. Since the Legislation Act 1988 prohibited the use of legally binding agreements to have contractors employ local residents, the focus of most of the Challenges was to use voluntary codes and charters to employ local people which also had parallel complementary training initiatives.[3] The following table (Table 8.2) shows the targets for jobs and training and the achievement numbers. As can be observed from the numbers, the job placement targets were exceeded while the training targets were not met.

Table 8.2. Construction initiatives job and training placement targets and achievements 1993/94 and 1994/95.[1,2]

	1993/94		1994/95	
	Jobs	Training places	Jobs	Training places
Target	774	5075	978	5548
Attainment	1636	3030	1184	2237
Attainment as percentage of target	211	60	121	40

Notes:

[1] No targets were set in 1992/93; attainment in that year was 115 jobs and 108 training places.
[2] Based upon information from 20 Challenges.

Source: Hayton 1998.

The main challenges in getting people for training include: small target areas; construction industry being also very skilled and highly competitive; training places were difficult to get because there was no financial incentives for employers to provide them; and in some cases, local residents had very low level of skills, maybe because they had been out of work for a long time. As can be seen from Table 8.3 below, highly skilled labour was employed in most positions:

Table 8.3. Skill levels of those placed into work through City Challenges construction initiatives—1992/93 to 1994/95.

Skill level	Job placement	
	Number	Percentage
Skilled manual	1585	54
Semi-skilled manual	675	23
Unskilled manual	643	22
Administrative/clerical	32	1
TOTAL	2935	100

Source: Hayton 1998.

---

[3] Most of the times the motivation was the persuasive abilities of those running the initiative and the self-interest of contractors, as the people running the initiatives were also responsible for awarding the future contracts.

The cost of achieving these placements and training was relatively modest, averaging to a total of £ 117, 945 per initiative in 1993/94, an 11% increase to £ 131,295 in 1994/95 and an estimated £ 139, 840 in 1995/96. The average cost per positive outcome was very low: £ 356 (in 1993/94) and £ 543 (in 1995/95) as compared to other initiatives involving training and gainful employment. This is what has made City Challenges quite attractive in spite of some of its shortcomings.

'New Economy' Theory Framework—Relationship between Economy and Culture after Regeneration

With regeneration generally transformation of the place is achieved—changes are structural, cultural and job opportunities are made available. However, as mentioned in the Cities of Civilization (Hall 1998) the exact form of transformation is dependent on the historical and geographical features of the city itself, and on the local leadership, entrepreneurship and talent. This could be seen way back in the Italian Renaissance (competition between Florence and Siena) and even more recently in Canada (Toronto and Montreal) or in the United States (Los Angeles and San Francisco).

The new economy does not exist on its own, it is woven in the urban fabric where it is found integrated within the concrete itself. The two well-known theorists of urban new economy are Allen Scoot and Richard Florida. Both of them recognize that place is central in understanding the urban new economy. For example, Scott (2000, p. 319) "insist[s] above all" on the centrality of "synergies that lie at the intersection between agglomeration processes [of the new economy] … and the cultural meaning of place". Scott (1988) focuses on post-Fordism, with an emphasis on the new industrial economy of 1980s "that emerged during that decade were highly specialized vertically disintegrated firms that produced niche products, operated within unstable markers, were linked by tight-knit, external exchanges through value-added networks and utilized specific types of labor that could be temporary and shifting, but in some cases also highly skilled" (Barnes and Hutton 2009). Factors such as market instability, density of transactions, role of institutions and need for skilled labour the firms spatially agglomerated and "triggered a reciprocal process between the flexible specialised industrial complex and the place it occupies" creating synergies between the production complex and the place. Scott then argues that cultural products industry such as entertainment complex, clothing, furniture design companies, printing and publishing meet this criteria of new economy (Scott 1996, 2000). While Florida (2002, p. 6) mentions, "place has become the central organizing unit of our time, taking on many of the functions that used to be played by firms and other organizations". Florida lays less emphasis on the internal structure and more on the creativity of the labour force. According to Florida (2002, p. 4) the main driving force in our economy and society is the creativity of the people working. He (Florida

2002, p. 223) further writes that the places of the new economy are set "by the location choices of creative people—the holders of creative capital". For Florida, the right qualities include:

What's there: the combination of the built environment and the natural environment …

Who's there: the diverse kinds of people, interacting and providing cues that anyone can plug into and make a life in the community.

What's going on: the vibrancy of street life, café culture, arts, music and people engaging in outdoor activities (Florida 2002, p. 232) (as mentioned in Barnes and Hutton 2008, 1250 and 1251). Barnes and Hutton (2008) argue that both the macro- and the micro-geographical specificities as well as contingencies are central to understanding the urban new economy.

Case Studies

The three case studies below illustrate examples at different levels:

1. Successful regeneration activity carried out in the UK
2. A plan for regeneration activity for Elliot Lake in Canada
3. Income diversification plan for Kuwait.

Case Study 1: Hartlepool Borough Council, U.K.[4]

Context

In 2001, unemployment in Hartlepool was 8.2%, which was above both the regional and national averages, 6.6 and 3.5% respectively. Hartlepool also had a very high level of multiple deprivations with 56% of the borough's population living in the 10% most deprived wards in England. Hartlepool Borough Council's Economic Development service developed a vision for regeneration and improving job opportunities in the region.

Vision for Hartlepool

The corporate vision of Hartlepool Borough Council is...

to work with others to support and develop a vigorous and diverse local economy and to get Hartlepool people into jobs.

and it addresses two of the key aspects of economic regeneration: a strong, diverse local economy and quality jobs and decent income for all.

---

[4] Audit Commission, nd.

Key Aspects of Economic Regeneration

(A) Creating the Conditions for Job Growth

Hartlepool encouraged job growth by focusing on a series of aims that reflected the building blocks that make up a strong, diverse economy. This was achieved by:

- ensuring availability of good choice of sites and buildings for businesses: the city itself got involved in managing Brougham Enterprise Venter and also helped in improving the industrial estates and refurbishment of the Oakesway Industrial Estate
- encouraged inward investment: redevelopment of the Marina and the town centre as well as securing finance for property refurbishments encouraged inward investment
- the council also supported local businesses by offering advice and support: The economic development service worked closely with other business support agencies, (such as Business Link Tees Valley, the Hartlepool Enterprise Agency and City Centre Training) to provide help to existing businesses to help them to grow through financial assistance, advice and guidance in partnership with other agencies
- and supported development of the tourist sector.

(B) Ensuring that Local Hartlepool Residents Get the Jobs Created

The Council started a range of programs that linked local people to available job opportunities created by regeneration. Efforts were made to consistently improve schools, investment in a community-based adult education network and a well-managed further education college.

The service had four major schemes:

- *Jobs Build*—the initiatives under this scheme include subsidies to employers in order to increase employment and training opportunities for the residents of a Single Regeneration Budget area.
- *Targeted training*—this includes short courses that are tailored to the recruitment needs of employers or to areas of labour market demand.
- *Women's opportunities*—provides support for women seeking to re-enter the labour market via training, employment or self-employment.
- *Intermediate labour market*—provides work-related training in order to get long-term unemployed people back to work.

Case Study 2: Proposed Economic Renewal Strategy for Elliot Lake Region[5]

Economic Renewal activities in Elliot Lake focused on: quantitative targets, on developing competitive fundamentals, increase value added activities (move away from exclusive dependency on one dominant activity; develop a stronger presence in the new economy; emphasize renewable sources of income; develop technological capabilities; involve higher education institutions; attract and retain a creative pool of talent) and participation from all.

Issues: While it may be convenient to argue that Elliot Lake economic difficulties in the past can be explained totally by the decline in mining activity, the truth perhaps lies elsewhere. The fact that mining activity alone could have affected so adversely all economic indicators of performance of the City is itself revealing. In this respect the heavy dependence on mining uranium is symptomatic of the general economic malaise in Elliot Lake and in many northern resource dependent communities.

The economy of Elliot Lake today remains generally undiversified but surely to a lesser extent than it was in the 1970s. Few activities remain as the exclusive economic engines of the City. Manufacturing activity is quite limited, disarticulated, traditional, inward looking, and technologically dependent on outside sources and control. Limited technological capabilities are developed or employed within the City, support services for businesses are scarce and entry into the knowledge economy is spotty and rare. The City cannot afford to gamble on 'sun set' industries and old Fordist and smokestack manufacturing activities. There is considerable evidence that success in the new globalized world of today is rooted in sun-rise activities of the new economy.

The current economic problems of the City are serious and there is little that a renewal strategy framework can do to deal with them individually or collectively. But an appreciation of what went wrong and that simplistic and borrowed solutions will not suffice is critical for reversing the negative operating mechanisms of the economy and for building the foundations for tomorrow's growth. It is generally believed that a correct diagnosis of the problems and challenges represents 70% of the solution.

There is also a general tendency in the City as elsewhere to underestimate the positive achievements and to exaggerate the negative trends. The City has been successful in arresting decline and has developed a few growth poles against formidable odds. There is still more to be done. What must be avoided are simplistic solutions that tend to exaggerate the implications of one paradigm or another. Most of the tendered recommendations these days involve Neo-Schumpeterian solutions and/or the perpetuation of the mythology of technological solutions. A more balanced, nuanced and eclectic

---

[5] Provided by Dr. AtifKubursi.

perspective is needed. Below are some of the ingredients of what might work for Elliot Lake.

Renewal Strategy

The City needs a renewal strategy, part of a social project involving its entire population. First and foremost there is a critical need for well-defined quantitative targets with pre-determined milestones. One such simple target is to raise the per capita income in the City to the provincial level. It is an easy target to monitor progress towards and very specific measures can be taken to govern its achievement. A few initiatives can help realizing this target. Below are a few prescriptive measures:

The City needs to move to productive activities with high value added. This will increasingly depend on building innovation capabilities, entrepreneurial and technical skills, appropriate educational and research centres, the requisite infrastructure, and a full-fledged and deliberate entry into the new economy. To move into high value added and sustainable activities the City must build and strengthen its competitive fundamentals. These call for:

- Empowerment of the business sector and community initiatives through strategic partnerships and by providing the enabling environment to sustain and solidify their participation.
- Adoption of effective policies and promoting institution building to encourage and develop domestic initiatives and co-ordinate the collective community effort to attract more investment.
- Raising and retaining skill levels. The emphasis here should be on building domestic productive capacities that can absorb and train on the shop floor, in the schools and everywhere, the skills of the residents.
- Increasing domestic technological capabilities by attracting and expanding higher education institutions.
- Developing linkages and networks between local businesses and non-local businesses that should be seen as techno-economic laboratories and agents of knowledge creation and dissemination. This is accomplished by helping firms strengthen their internal problem-solving capacities (through skill upgrading and building competencies) and through fostering external linkages to other firms and knowledge producing institutions.
- Increasing the share of the new economy and the knowledge economy. Competitiveness now seems to depend on getting the right information and knowledge to the right place at the right time. This needs a viable and efficient informational infrastructure from Internet connections, to web sites, to satellites, to fibre optics, to governments opening up the information highway corridors.
- Continuous tapping into local talents and dismantling any barriers that preclude the full participation of residents, young and old.

Focusing on the competitive fundamentals and increasing value added activities that are divorced from natural resource dependency have implications for change throughout the economy—for business, for other sectors, for local, provincial and federal governments, for regional institutions and for the economy and society as a whole. A consistent picture has recently emerged from the diverse literature on technological gaps, information gaps and knowledge gaps. The potential for 'catch-up' is there, but is only realized by cities that have a sufficiently strong 'social capability', e.g., those that manage to mobilize the necessary resources (investments, education, R&D, etc.) and actors (people, firms, entrepreneurs, government, universities, unions, etc.). These factors should also be seen as complements rather than as substitutes in economic growth.

It is equally crucial to not treat technology as 'blueprints' or 'designs' that can be bought and sought in the market. Rather it should be treated as organizationally embedded, tacit and cumulative in character, influenced by the interaction between firms and their environments, and geographically localized.

Governments will still have to act in critical areas for catch-up to happen and to have the greatest benefits in strengthening the City's competitive fundamentals:

- Changing the way the City invests for the future: Putting strong emphasis on investing in people, training, information and knowledge.
- Changing the way the public sector relates to the private sector: Emphasizing the development of sectoral strategies, strategic groups of companies, community initiatives and local/regional. Above all removing the impediments on the full participation of residents in all aspects of development.
- Changing the management of the economic change: Finding winners, building on strength and creating flexible systems for a more adaptable economy.
- Building the needed infrastructure in all of its aspects—the physical, the informational, the organizational and technological.

Although the City Administration has an important and vital role to play in this economic renewal process, it cannot alone make the policy work. Everyone must work together to develop an economy with built-in capacity to upgrade and continuously move to higher value added and to the newer realms of the economy. Tapping into the federal programs, provincial initiatives and private Foundations is necessary and critical.

The renewal strategy we are defining is not a budget, or a short-term stabilization policy or even a plan. It is a framework that is intended to create a shared vision and a common sense of direction that shapes the way all segments of the economy and society can work together.

Case Study 3: Kuwait—Diversification of Income[6]

The idea behind elaborating this case study is to emphasize the need for economic diversification that does not depend on just one source of income. This becomes more important when planning regeneration activities to broaden the economic base of the country. The example below can be used to see how diversification can be achieved.

Context

The economy of Kuwait is highly dependent on oil and in the recent past, it has not achieved any significant progress in reducing its heavy dependence on oil revenues, and diversifying its productive capacities and export basket. Consequently, the economy is becoming more vulnerable to external shocks, due to the prevalent volatility of world demand and prices of oil that ultimately affect the growth of GDP, public revenues and the government's capacity to finance its ever increasing current expenditures and consequently the welfare of the population.

Over the past decades, the weight of the private sector in the economy has dropped and its contribution to exports, employment, and the generation of wealth has declined. The government must revise its development strategy in order to reduce pressure on public finances and to offer jobs to nationals in the private sector. Energizing the private sector is believed to be the best option to foster economic diversification and transformation of the country.

International experience and empirical evidence suggest that private sector development contributes to the economic transformation of the economy through the diversification of its productive structure and exports' base. Energizing the private sector strategically can address the challenges of export promotion and bring about the development of a basket of goods based on comparative advantage that can simultaneously increase the capacity for job creation.

In Kuwait, the unintended consequence of the government's benevolent and protective philosophy has marginalized the private sector causing it to be uncompetitive because: (i) the enabling environment is over-regulated; (ii) the public sector has crowded out the private sector in productive activities, especially those where there are clear comparative cost advantages (e.g., downstream oil activities); and (iii) the overall regime of subsidies has created pervasive price distortions (World Bank 2001). As a consequence, the private sector in Kuwait is very small. In many of the critical activities in the Kuwaiti economy, including the oil sector, electricity, gas and water, transport and telecommunications, the share of the private sector is miniscule if not completely absent. The picture seems even more serious when one

---

[6] Provided by Dr. AtifKubursi.

considers employment demographics where approximately 95% of nationals are employed in the public sector.

A large number of economic problems can arise as a result of a small private sector. At the macro-economic level, the government may be unable to finance its expenditures in the long term when oil revenues drop, or if revenues are primarily generated through the profits of publicly owned companies, or it may do so by distorting relative prices and exploiting monopoly power. The enterprise sector may be inefficient and not internationally competitive because of inadequate managerial incentives and a poor corporate governance structure. An insufficiently competitive market environment may also encourage inefficiency, and a stifling bureaucracy may act to limit the formation of new firms, especially small and medium-sized ones (SMEs). It is often argued that the latter deficiency may be especially serious for job creation and innovation. Moreover, the prevalence of a host of political economic problems can arise whereby bureaucrats develop additional rules and regulations for the private sector, not to further the economy, but instead to create opportunities to further their own political and economic objectives. Based on international experiences, the development of the private sector has been achieved through three general channels: (i) boosting private and Foreign Direct Investment (FDI); (ii) accelerating privatization of inefficient public enterprises; and (iii) developing domestic Small and Medium sized Enterprise (SME). However, the ability of a country to make progress in implementing the above three strategies is conditioned by the structure of the economy itself. In particular, government policies and instruments that can present strong obstacles towards the development of the private sector in general and to the inflows of FDI in particular should be identified and removed. For example, foreign investors are not allowed to operate in some sectors of the economy while a package of obstacles harms their implementation in many others. The second determinant is the macroeconomic situation of the country itself. Progress on privatization is very different in a country that enjoys structural budget surpluses versus a country that is facing structural public deficits. The last determinant is the institutional environment, which includes the labour market.

To boost the contribution of the private sector in the Kuwaiti economy, a package of reforms could be formulated and suggested to the government. However, prior to moving to recommendations, it is important to answer a fundamental question: which sectors should Kuwait focus on in its diversification strategy? Compact Product Spaces developed by Hidalgo (2006, 2007, 2009, 2009a, 2009b, 2011, 2011a) and Hausmann and Klinger (2007) offer glimpses of hope and a possible diversification strategy. This approach builds on revealed comparative advantage coefficients and suggests that success of diversification depends on capitalizing on proximate competencies. Entrepreneurs are more comfortable with activities with which they are familiar and can transfer their skills, experience and knowledge to them.

Conclusion

As discussed in the chapter it is important to base urban regeneration planning on a three-pillar approach so that the local community has both socio and economic benefits from the intervention. The chapter also described how to put value on both direct and indirect benefits of regeneration activities along with creating jobs during regeneration and after regeneration. Successful creation of good jobs for local residents usually involves subsidized or complementary training by the local governments and a host of supporting programs.

References

Adair, A., Jim Berry, Stanley McGreal, Norman Hutchison, Craig Watkins and Kenneth Gibb. 2003. Urban regeneration and property investment performance. Journal of Property Research 20(4): 371–386.

Adair, Alastair, Norman Hutchison, Jim Burgess and Stephen Roulac. 2005. The appraisal of urban regeneration land: A contemporary perspective allowing for uncertainty. Journal of Property Investment and Finance 23(3): 213–233.

Audit Commission. Nd. Economic and community regeneration: Learning from inspection. Local Government Briefing. Accessed from www.audit-commission.gov.uk.

Barnes, T. and Thomas Hutton. 2009. Situating the new economy: Contingencies of regeneration and dislocation in Vancouver's Inner City. Urban Studies 46(5&6): 1247–1269.

Bond, Sandy. 2001. Stigma assessment: The case of a remediated contaminated site. Journal of Property Investment and Finance 19(2): 188–210.

Florida, R. 2002. The Rise of the Creative Class: And How is it Transforming Work, Leisure and Everyday life. Basic Books, New York.

Hall, P.G. 1998. Cities in Civilization. Weidenfeld and Nicolson, London.

Hausmann, R. and B. Klinger. 2007. Structural Transformation in Chile. Quantum Advisory Group. Working Paper.

Hayton, Keith. 1998. Using construction projects to create jobs and training opportunities: The City Challenge Experience. The Town Planning Review 69(2): 115–134.

Hidalgo, C.A. 2006. The Product Space and Its Consequences for Economic Growth. Presentation at the LIEP talk series, Kennedy School of Government, Harvard University. http://www.chidalgo.com/Talks/ProductSpaceAPS.ppt.

Hidalgo, C.A. 2007. The Product Space Conditions the Development of Nations. Presentation at NetSci07. http://www.chidalgo.com/Talks/ProductSpaceSFI_07.ppt.

Hidalgo, C.A. 2009. The Dynamics of Economic Complexity and the Product Space over a 42 year period. CID Working Paper No. 189.

Hidalgo, C.A. 2009a. The Product Space and Economic Complexity. Presentation at ISCV, Valparaiso, Chile. http://www.chidalgo.com/Talks/ProductSpace_And_ EconomicComplexity.ppt.

Hidalgo, C.A. 2009b. The Product Space and The Building Blocks of Economic Complexity. Presentation at an Asian Development Bank Meeting in Almaty, Kazakhastan. http://www.chidalgo.com/Talks/CA_Hidalgo_Almaty_09.ppt.

Hidalgo, C.A. 2011. Discovering Southern and East Africa's Industrial Opportunities. The German Marshall Fund of the United States, Economic Policy Paper Series 2011.

Hidalgo, C.A. 2011a. The Product Space and Its Consequences for Economic Growth. Presentation made in the Center for International Development—Kennedy School of Government—Harvard University. http://www.chidalgo.com/Talks/ProductSpaceAPS.ppt.

HM Treasury. 2008. The Green Book. Appraisal and Evaluation in Central Government. Treasury Guidance. London. TSO. www.hm-treasury.gov.uk/green_book_complete.pdf.

ODPM. 2004. Assessing the impact of spatial interventions. Regeneration, renewal and regional development. 'The 3Rs guidance'. London: Office of the Deputy Prime Minister.

Otsuka, Norkika and Alan Reeve. 2007. The contribution and potential of town centre management for regeneration. Shifting its focus from "management" to "regeneration". TPR 78(2): 225–250.

Roberts, P. 2000. The evolution, definition and purpose of urban regeneration. pp. 9–36. *In*: P. Roberts and H. Sykes (eds.). Urban Regeneration: A Handbook. SAGE Publications, London.

Scott, A.J. 1988. Metropolis: From the Division of Labor to Urban Form. University of California Press, Berkeley, CA.

Scott, A.J. 1996. The craft, products and cultural products industries of Los Angeles: comparative dynamics and policy dilemmas in a multisectoral image-producing complex. Annals of the Association of American Geographers 86: 306–323.

Scott, A.J. 2000. The Cultural Economy of Cities: Essays of the Geography of Image Producing Industries. Sage, London.

Tyler, Peter, Colin Wamock and Allan Provins with Peter Wells, Angela Brennan, Ian Cole, Jan Gilbertson, Tony Gore, Richard Crisp, Anne Green, Mike May-Gillings and Zara Phand. 2010. Valuing the Benefits of Regeneration: Economics Paper 7—Logic chains and Literature Review. Department of Communities and Local Government.

Index